The Biosphere and Civilization: In the Throes of a Global Crisis

Victor I. Danilov-Danil'yan • Igor E. Reyf

The Biosphere and Civilization: In the Throes of a Global Crisis

Translated from Russian
by Steven McGrath

 Springer

Victor I. Danilov-Danil'yan
Correspondent Member of Russian
Academy of Sciences
Director of the Water Institute
at the Russian Academy of Sciences
Moscow, Russia

Igor E. Reyf
Freelance Journalist/Writer
Frankfurt am Main, Germany

ISBN 978-3-319-67192-5   ISBN 978-3-319-67193-2   (eBook)
https://doi.org/10.1007/978-3-319-67193-2

Library of Congress Control Number: 2018931891

Printed on acid-free paper

This Springer imprint is published by the registered company Springer International Publishing AG
part of Springer Nature.
The registered company address is: Gewerbestrasse 11, 6330 Cham, Switzerland

Dedicated to the memory of the remarkable Russian writer and advocate of popular science Rafail Nudelman.

Acknowledgments

Every book has its backstory, when the question is decided: is this going to happen or not? Unfortunately, the people involved in this fateful moment typically remain behind the scenes and are not placed in the book's imprint. And so, the only chance to give them their due is to name them here as an expression of gratitude.

First, we would like to thank Dr. Christian Witschel, who stood at the birthing bed of our first English-language edition, *Sustainable Development and the Limitation of Growth: Future Prospects for World Civilization* (2009). About 10 years ago, he believed in our new project and recommended it to the editors of Springer Nature. We'd like to thank the translator of this first English-language book, Dr. Vladimir Tumanov, and his colleague Mika Tubinshlak, who provided invaluable help in translating the publishing application with all its corresponding materials. Without this help, the decision to publish this book could not have been made.

We were also fortunate at the next stage, when our application fell into the hands of Dr. Sherestha Saini, Senior Editor for Environmental Science at Springer Nature. I have crossed paths with no small number of editors and publishers, so I have room to make comparisons. I wish that every author could meet such a sympathetic and understanding editor as Dr. Saini, whose cooperation over the previous year has left me with only fond memories.

And finally, there was the long and difficult process of translating the book from Russian into English, as a result of which it now speaks with a new voice. I was, of course, aware of how difficult a task stood before him, and, naturally, I was worried. But my worries proved groundless, and Steven McGrath managed the task wonderfully, even though this was, seemingly, his first such experience. Here is how one specialist in literary translation responded to his work in a letter to me: "I think you are in very good hands, i.e. you are truly in luck. He has a wonderful command of scientific language. His style corresponds to yours. He's not just translating—he's clearly thinking." Indeed, we truly found in Steve not only a good translator, but a like-minded person, and that, I think, is the greatest insurance of success.

To conclude, I'd like to thank three people close to me whose contributions, perhaps, are not as obvious, but without whose help and sympathy it would have been much more difficult for me to reach this goal: my wife Zaytuna Aretkulova, my son and my constant consultant Vitali Reyf whose fluency in English helped me to overcome the language barrier, and Bella Leibovica.

I would also like to express my appreciation to our reviewers, who gave this book the green light before its translation into English. I look upon their positive report as an advance given us, and I'd like to think that, when they open this book, it will not disappoint.

Igor E. Reyf

Preface

Who has never heard of the ecological crisis, the ecological *challenge*, facing us in our time? Many see it as the most important challenge of modern civilization. Which challenge is unimportant? Aren't the terrorist attacks of September 11th important? Or the mounting waves of refugees pouring from North Africa and the Middle East into Europe? Does the reader believe that we, the authors, seek to trivialize any issue that stands beyond our ecological ken?

Of course not. But the systemic ecological approach at the heart of this work will allow you to look with new eyes upon the critical situation of modern civilization as a whole. You might just see, hidden in the depths, the links between these seemingly unconnected phenomena. That the crisis has reached a perilous stage is apparent even to the untrained eye. Perhaps the outbreak of global terrorism, declaring itself loudest of all among modern challenges, is but one symptom of a common malady. It comes alongside catastrophic changes to the environment, explosive population growth in developing countries, large-scale technological disasters, and epidemics of previously unknown infections. All of these challenges, often called the challenges of globalization, do not limit themselves to individual peoples and states, but threaten all of humanity.

The idea of a "challenge" as a historical and philosophical concept was coined in the middle of the last century by English historian Arnold J. Toynbee in his famed multivolume work, *A Study of History*. As a Christian thinker, Toynbee understood this to mean an ongoing dialogue between mankind and Divine Reason (*Logos*), by which people would ultimately realize their true nature and greater historical destiny. Each test of strength, whether by nature or by rival tribes, would, according to Toynbee, serve as an engine of historical progress, bringing forth the creative energy of the nation and raising it to a new level of development. Sometimes, overcoming the challenge would allow the birth of a new civilization. A challenge prompts growth, Toynbee argues. In its response to the challenge, society solves a basic life problem that it faces, which allows it to move to a higher and more advanced stage of development (McNeill 1989).

Even with the eternal stretch of "challenge-and-response" going back to hoary antiquity, the current ecological challenge is unique. This is because, for the first time in millennia, the question is before all humanity—before humans as a species—"To be, or not to be?" We're not talking about an asteroid on a collision path with Earth, which is unlikely to come in tens of thousands or even in millions of years. We are talking about the day-to-day process of degradation to our environment, destroyed by human economic activity, which has reached the critically dangerous limit of its historical course.

Do the seven billion inhabitants of earth know about this? Not the bona fide ecologists, but the average residents, or as they say, the man on the street (which includes, we're sorry to say, most politicians and the lion's share of our business and cultural elite), on whom the fate of future generations depends?

Yes and no. Yes, because apocalyptic warnings trickle through in print and broadcast media concerning the ramping up of the greenhouse effect, the expanding hole in the ozone layer, and the pitiless butchery of "the lungs of the planet," the tropical rainforest. No, because people's day-to-day consciousness has the remarkable ability to wander away from such information, into quaint imaginings of a distant and ever-receding deadline, or to hide from the ecological threat within the technological cocoon of modern civilization.

Against this backdrop, can ecologists realistically hope to reach people? After all, only a narrow circle of experts has access to and can understand the specialized research material. The forecasts and conclusions that come from the research, however, are addressed to those without any connection to the science of ecology, though, it is these nonspecialists who will ultimately determine the effectiveness or ineffectiveness of information which carries dire implications for the fate of the world.

And yet we would argue that there have been precedents, or at least analogous situations in history. In 1939, at the urging of his European colleagues, Albert Einstein wrote a letter to President Franklin Roosevelt in order to convince him of the need to start full-scale work on a nuclear program. Nothing was certain at the time. There was no reliable information, only guesses and suppositions that Nazi Germany was conducting similar research. No physicist could guarantee that the chain reaction to splitting uranium-235 atoms observed under laboratory conditions would truly lead to the predicted nuclear blast. And the physicists themselves had little influence, being known only in narrow academic circles.

A great deal weighed upon that moment, but the decision was not for the scientists to make. That responsibility belonged to the president and his administration, who had no particular knowledge of nuclear physics at all. Nonetheless, the fateful decision was made. That decision's influence, for the better (the nuclear deterrent factor) and the worse (the ruinous arms race, the recent threat of nuclear terrorism), had consequences far beyond the immediate concerns of the Second World War.

Why have we recalled this particular case? Because the position of ecological scientists today resembles the situation facing atomic physicists in the late 1930s. They too cannot yet show the world the "bomb" ticking away in the foundations of modern civilization. Their predictions and estimates are based not as much on known precedent as on the logic of ongoing processes in the biosphere which promise to cause (or have already caused) irreversible change.

The problem is that when the change becomes apparent to the public at large, the time for taking appropriate measures will have already passed. Only trust in the knowledge and foresight that comes from science (as with the American atomic project) can serve as a reliable basis for preventing ecological catastrophe. In that sense, scientists now have a greater responsibility than ever before. Come to think of it, so too do global political, cultural, and business elites, as well as those often referred to as concerned citizens. And, if we may say so, there is no more important task than working together.

Moscow, Russia   Victor I. Danilov-Danil'yan
Germany   Igor E. Reyf

Reference

1. McNeill William H. (1989) Arnold J. Toynbee: A Life. — New York: Oxford University Press.

Contents

Part I
Civilization in Crisis: The Edge of the Abyss

Chapter 1
The Global Ecological Situation

"Thirty years ago," zoologist Viktor Dolnik wrote in 1992, "only a few ecologists on the whole planet thought about the approaching ecological catastrophe. The public called them alarmists and had a big laugh at their expense. Today, though, large numbers of ordinary people have felt for themselves the growing pressure of primary factors (affecting human life)"[1] (Dolnik 1992).

Indeed, people have come to think ecologically at a rate unusual by historical standards. The topic frequently appears on television and online. Magazines dealing entirely with ecological problems come out one after the other. Impressive international conferences regularly gather to discuss environmental protection at the highest levels. In 1972, the United Nations formed a permanent body for the issue, the UN Environmental Program (UNEP). The UN Commission on Sustainable Development, a functional commission of the Economic and Social Council, arrived 20 years later with the aim of implementing the international agreements on environmental issues reached at the 1992 Rio de Janeiro Earth Summit. Aside from these, authoritative non-governmental organizations such as the World Wildlife Fund (WWF) and the Global Footprint Network have begun work in most countries.

Ecology has also broken into the worlds of business and politics. By 2010, the market for green technology surpassed the $1 trillion mark. Political party platforms can no longer do without promises to fix one environmental problem or another. Green parties have gained representation not only in European parliaments, but in cabinets (from 1999 to 2005 in Germany, for example), directly influencing government programs and funding nature-friendly projects. Finally, we should recall that in 2007 the Nobel Peace Prize was awarded to former Vice-President Albert Gore and the Intergovernmental Panel on Climate Change (IPCC), "For their efforts to build up and disseminate greater knowledge about man-made climate change, and to lay the foundations for the measures that are needed to counteract such change" (Nobel Prize 2007).

---

[1] Parentheses ours.

© Springer International Publishing AG, part of Springer Nature 2018
V. I. Danilov-Danil'yan, I. E. Reyf, *The Biosphere and Civilization: In the Throes of a Global Crisis*, https://doi.org/10.1007/978-3-319-67193-2_1

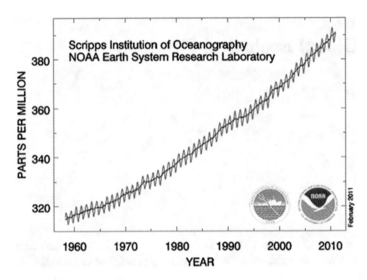

Fig. 1.1 A graph of the increase in CO_2 concentrations in the atmosphere (parts per million), taken at the Mauna Loa Observatory, Hawaii, from 1958 to the present. Slight fluctuations in the course of the overall trend reflect seasonal variations in CO_2 levels connected to intensified photosynthesis and carbon use by vegetation in spring and summer. Source: Scripps Institution of Oceanography NOAA Earth System Research Laboratory

It would seem that all the necessary financial and technological resources have been mobilized. But the problem, like a giant iceberg, is still sitting right in the path of global civilization, and it shows no signs of melting. Meanwhile, people are gradually learning to think of "the environment" as a long-term problem, one that their children and grandchildren will live with. They've learned to think that the relatively carefree days of the recent past are never coming back, and that mankind can go on living with the current troubles (sometimes better, sometimes worse) forever, if need be.

In reality, the ecological situation we are living through is markedly different from anything the human race has dealt with before. If for no other reason, this is because the dangerous changes have taken on a global character. They have spread to every subsystem and component of the environment. They have reached the entirety of the planet's surface from pole to pole as various scientific studies have confirmed, perhaps sparing only the ocean depths.

Particularly telling is the concentration of **nutrients**—substances that take part in life processes—in the atmosphere. Studies of air bubbles in glacial core samples from Antarctica and Greenland, which keep a record of the atmosphere in long-past epochs, have shown that concentrations of nutrients are changing faster than at any time in hundreds of thousands of years at the least (Barnola et al. 1991; Cannariato et al. 1999) Most of all, this concerns the increase in the concentration of atmospheric carbon dioxide (CO_2).

Since 1958, when consistent monitoring began, the concentration of CO_2 in the atmosphere grew from 315 to 390 parts per million (ppm). (See Fig. 1.1.) At the same time, ice cores from the Vostok Antarctic Research Station show that over the last four ice age cycles (about 400 thousand years), CO_2 levels varied from 190 ppm during glaciation to 280 ppm during interglacial periods (Rapp 2008). During that period, the rate of carbon level increase was lower by two orders of magnitude, while the decrease from peak to trough took up roughly 10,000 years.

A 3 km bore conducted by the European Project for Ice Coring in Antarctica from 1996 to 2006 has allowed us to glimpse an even more distant past, going back 800–850 thousand years. As University of Bern Climatologist Thomas Stocker notes, in the entire period recorded in the core carbon dioxide levels never once rose above 290 ppm. It was only with the approach of the present day that concentrations of CO_2 began rising sharply. In the past 50 years, the rate of increase has surpassed anything in the observed ice record by 200 times(!) (Siegenthaler et al. 2005). Analysis of the ratios of Carbon-14 and Carbon-13 isotopes in atmospheric CO_2 demonstrates with a high degree of certainty that the origin of the increase is connected to fossil fuel combustion and other human economic activity (Vitousek 1994).

Granted, coal was known as early as ancient Rome, but until the mid-nineteenth century, wood, straw and charcoal served most of humanity's energy needs. Only after that point did fossil fuels replace them as a primary source of energy. We trace the skyrocketing increase in CO_2 emissions to that moment, with the process accelerating in the last century. The emissions come from non-industrial as well as industrial sources such as cement production and gas burn-off from oil drilling. They are growing ever faster. The growth rate for CO_2 emissions rose from 1.0% in 1990 to 3.4% in 2008, more than tripling nine billion metric tons per year (Le Quere et al. 2013). The quickly developing economies of China, India and Brazil made up most of that difference, along with the growth of the global automobile park (Oak Ridge National Laboratory 2011).

Unfortunately, fossil fuel carbon emissions continued racing higher into the twenty-first century, reaching about nine billion metric tons (nearly ten billion standard tons)/year in 2008. For this we must thank the quickly developing economies of China, India and Brazil along with the world's ever-growing auto park (Oak Ridge National Laboratory 2011).

By now every grade-schooler probably knows that carbon dioxide plays a major role in what we call **the greenhouse effect**. Less well known is that the greenhouse effect provides just as much support for the conditions of life on earth as the atmosphere itself. Greenhouse gasses "capture" part of the sunlight reflected by the Earth's surface, warming the lower levels of the atmosphere. This results in a roughly 30 °C increase to the surface temperature. So, the greenhouse effect itself does not present a danger, but rather exceeding its baseline level, which has remained unchanged for hundreds of millions of years. Think of it as too much of a good thing.

True, climatologists disagree on the share of human contribution to the global warming confirmed in countless observations over the twentieth Century (Kondratyev and Donchenko 1999; Jaworowski 1997). However, the first decade of the twenty-first century was the warmest on modern meteorological record, and summer of 2015 turned out hotter than any other in the history of the northern hemisphere. The rate of warming was particularly significant in the 30 years from 1980 to 2010 (National Research Council 2011). Over the course of the twentieth century,

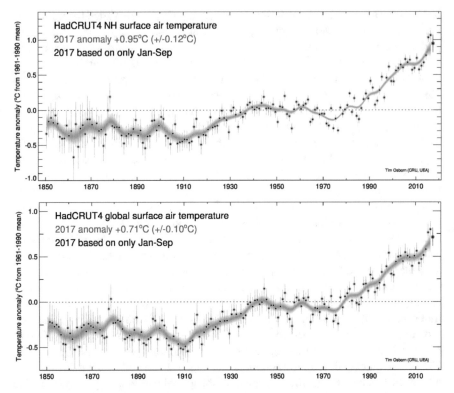

Fig. 1.2 Yearly Anomaly in near-surface temperature for the northern hemisphere (land) and globally from 1850 to 2017. Source: Climatic Research Unit, University of East Anglia https://crudata. uea.ac.uk/~timo/diag/tempdiag.htm

average surface temperatures rose 0.7 °C, surpassing fluctuations for the whole previous millennium (Fig. 1.2).

Of course, the rate of warming varies between regions of the globe. The highest rate is observed in continental areas at middle latitudes of the northern hemisphere. In eastern Siberia west of Lake Baikal, for example, mid-winter temperatures have risen by nearly 2 °C. Warming is less noticeable at oceanic middle latitudes and in the southern hemisphere. In a few areas of the Southern Ocean and Antarctica, we have even observed some cooling.

With this we cannot help but notice the correlation between the increase in surface temperature and the accumulation of carbon dioxide gas in the atmosphere over the course of the twentieth century (Fig. 1.3). While we can expect a slowed increase in the release of CO_2 into the atmosphere in the future as renewable sources of energy replace organic fuel, this is unlikely to happen in the next 20 years. Meanwhile, the thawing of polar icecaps and subarctic Siberian bogs encased in ancient permafrost threatens to further crank up the speed of climate change. The thaw, itself the result of warming, causes a chain reaction of secondary effects such as the release of methane from the melting of long-frozen soils or from gas hydrates in the ocean depths as the World Ocean's temperature rises.

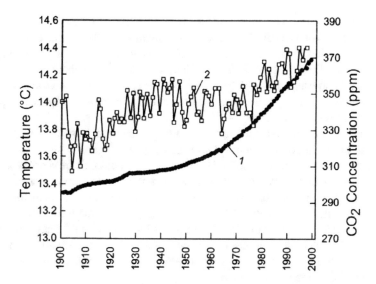

Fig. 1.3 Rate of change to the Earth's near-surface temperature (1) and the concentration of carbon dioxide in the atmosphere (2). Source: Worldwatch Database (2000)

The fact is that for thousands of years roughly 70 billion tons of methane, a fourth of the world's total, has been trapped under the ice of Siberian bogs. Until recently, these reserves were not taken into account in calculations of the rate of global warming, the assumption being that they would make themselves felt much later, when the climate had warmed. Over the past 10–15 years, however, as scientists have observed, the ongoing thaw of Siberian bogs has already become a fait accompli. That, it must be said, is one of the most unpleasant surprises that has awaited humanity due to global warming. The polar regions of western Siberia are warming faster than anywhere else on Earth, and methane's greenhouse effect is 20 times stronger than carbon dioxide's. The journal "New Scientist" quotes Professor Sergey Kirpotin of Tomsk University on this point: "[It is an] ecological landslide that is probably irreversible and is undoubtedly connected to climatic warming" (Pearce 2005).

In this way, assuming that the current rate of acceleration continues, global warming could reach 2 °C by 2060 and atmospheric CO_2 concentrations will surpass 1900 levels by 150% (Joshi et al. 2011; Rogelj et al. 2011). The consequences of these developments are obvious. It means a radical shift in the world climate zones. It means a rising sea level as continental ice sheets in Antarctica and Greenland melt and the World Ocean experiences thermal expansion. By the end of the twentieth century, sea levels were already rising by 2.1 mm/year, more than at any time in the past 2000 years (Kemp et al. 2011). It means the sinking of low-lying coastal territory, where nearly a third of the Earth's population lives. Finally, it means the transformation of the whole natural world, representing a threat to mankind's very survival.

But CO_2 is not the only or even the most important greenhouse gas (water vapor, for example, makes up an order of magnitude more of the atmosphere at 0.5–1%), and industrial pollution is just one source of its migration to the atmosphere. **Land use** plays no less of a role in this, having contributed 180 (198) trillion tons of atmospheric carbon between the Neolithic Revolution and the present day, while by the end of the twentieth century, industrial emissions added up to about 160 (176) trillion (Lashof and Ahuja 1990; Titlyanova 1994). This is because land use has caused *the destruction of ecosystems*, especially the cutting down of forests, which play a vital role in excess carbon fixation through photosynthesis.

Overall, the destruction and degradation of the ecosystem is, without a doubt, the largest and most important component of the global ecological crisis. Ancient agriculture served as the starting point for this process. Thousands of years before the industrial revolution began, the acquisition of new lands for farming was already leading to the destruction of enormous swaths of the natural biota. As historian Lev Gumilyov wrote, "Hard working farmers, thinking only of the next year's harvest, turned the banks of the Hotan and Lake Lop Nur into sand dunes.[2] They churned up the soil of the Sahara and let dust storms blow it away" (Gumilyov 2014). Worst of all, however, was the destruction of **forest ecosystems**, the most important stabilizing factor in the global environment.

The most crushing blow to ecosystems came in the twentieth century. While at the turn of that century, territories with ecosystems partially or entirely destroyed by man took up 20% of land, by the beginning of the twenty-first they occupied about 60% (not including ice-covered or denuded territory). In the meantime, three massive zones of environmental destabilization have formed in the northern hemisphere, covering a total area of 20 million km^2 (12.5 million sq. miles) (Arsky et al. 1997; Danilov-Danil'yan and Losev 2000; Nowinski et al. 2007). (For more on that, see Chap. 15.)

We will have plenty to say throughout this book about forests and their key role in nutrient cycles. Essential photosynthetic production takes place in forests. Among land ecosystems, forests have the greatest ability to absorb the excess carbon oxide gas thrown into the atmosphere during the combustion of fossil fuels. By storing and evaporating water, they provide most of the continental water cycle, support river flow, even out short-term and seasonal fluctuations, reduce the speed of pressurized air fronts that produce extreme weather, work as filters to clean the atmosphere, et cetera.

Currently, forests occupy about 40 million km^2, or 31% of land area. Before the Neolithic Revolution, 10,000 years ago, they had access to over 60 million km^2, or 45% of the land's surface (FAO 2010) Thus, in the course of history, humanity has annihilated no less than a third of the planet's forests. The Neolithic or Agricultural Revolution not only brought a start to farming culture, it also heralded a new stage in the relationship between human beings and nature. Their predecessors, hunter-gatherers, fit naturally into their environment, not unlike other species. Now they set out to conquer the world, acquire new lands and transform them for use by fields and herds.

The forests would have seemed to present a daunting hurdle for the new colonists to cross. But the slash-and-burn method of agriculture and new implements to fell

---

[2] A dried up river and closed-basin salt lake in western China's Xinjiang province.

trees successfully overcame the problem. True, these primitive and inefficient methods caused the plots to quickly exhaust themselves. This didn't worry the ancient farmer: without any shortage of land, he could move to a new plot whenever he pleased, clearing away the forest as he went.

The process followed a specific, geographically understandable pattern. First, the forests in the ancient civilization zones of India, China and the Near East were annihilated, followed by those around the Mediterranean in the millennium before the common era. The mass felling of European woods began later. Before the seventh century, they covered 75% of the continent. But with the Renaissance and the Age of Discovery, deforestation took on vast dimensions as cities boomed and nations built sailing fleets. Forests were cut to open tillage and pasture. People used wood as both fuel and raw material, for which the 1782 invention of the steam engine added still greater impetus. Meanwhile, the populations of European countries skyrocketed. As immigrants moved west in the eighteenth and nineteenth centuries, deforestation overtook North America as well.

As for deforestation on a global level, its rate and pattern reflected population growth until 1950. At that point, population growth increased sharply and caught up to deforestation, creating a kind of "scissors" shape (Fig. 1.4).

It's worth noting that population growth and deforestation reach their peaks simultaneously in the same regions. This partly coincides with the start of economic growth in a given country. Both rates then typically stabilize or slow once society reaches a certain level of prosperity. The fate of first the northern, then the southern forests illustrates this rule.

The *northern forest zone*, occupying roughly two billion ha (4.94 billion ac), lies mainly in three countries: Russia, Canada and the U.S. Peak eradication of these forests coincided with rapid industrial development in Europe and North America with its corresponding population boom and urban construction. It continued through the early 1900s. As a result, Europe lost the vast majority of its forests, which shrank to a mere 10% of its territory (State of the World's Forests 2012). Only as new technologies improved agricultural yields and food storage, and as new materials replaced lumber in construction and wood as fuel, did the process of deforestation wind down and a period of restorative forestry begin. And while forest coverage of Europe (not including Russia) is approaching 35%, this is, with minor exceptions, cultivated secondary forest and tree farms growing on the ashes and stump holes of dead ecosystems. They are at least four times less productive and biodiverse than primary forests, with which they cannot remotely compete as environmental stabilizers. Old-growth primary forests have hung on only in the mountainous Alps, Pyrenees, Carpathians and the Balkan Peninsula, along with northern Scandinavia and Finland. But one way or another, the main threats to European and North American forests have passed, except for global climate change.

Things stand entirely differently in the *southern forest zone*, where tropical forests have suffered an unprecedented assault since the 1920s. The following numbers will give you some idea of the scale of the assault. From 1990 to 2010, 88 million ha (217 mln ac) of primordial rain forest, 9% of the continent's total forest area, were cut down in South America alone. South America's rainforests shrank to less than

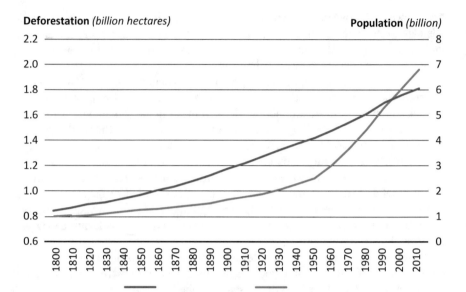

Fig. 1.4 Earth population and total deforestation 1800–2010. Verticals: on the left, deforestation in billions of hectares (ha) (1 hectare = 2.47 acres [ac]); on the right, population in billions. The lower line, forming a "scissors" with the upper, represents population growth. Source: State of the World's Forests (2012)

half of the continent's area for the first time in history. In Africa, where forests cover 23% of the surface, 10% of them were wiped out over the same period, a total of 75 million ha (185 mln ac). Some countries of the southern forest zone—El Salvador, Jamaica, Haiti—have lost their forests altogether. In nine countries forests are being annihilated at 2% per year, and in 20 more the rate of deforestation surpasses 1%. If this trend continues, many of these countries will lose their forests in the next century. At the very least, all will face serious ecological problems (State of the World's Forests 2012).

And so you might say that developing countries are repeating the unlearned lessons of industrialized states with a century's delay. As you can see from the diagram (Fig. 1.5), the latter passed this tragic baton to the former somewhere back in the mid-twentieth century. And while the pace of deforestation worldwide has recently slowed, the situation remains deeply troubling. According to the UN Food and Agriculture Organization (FAO), the area of world forests shrank by 13 million ha (32 mln ac) a year from 2000 to 2010 (primary forests shrank by 5.2 mln ha/12.8 mln ac). This is ten times faster than the process of natural forest recovery. The 130 million ha (321 mln ac) lost over the decade as a whole made up a full 3.2% of all forest areas from 2000 (FAO 2010; State of the World's Forests 2012).

Importantly, the reasons for developing countries' profligate use of forest bounty remain in force. These include inefficient agricultural systems in constant demand for new tillage and pasture. They include a lack of electrification and gas supply, which means that 100 million people depend on wood as their only source of fuel.

Million hectares

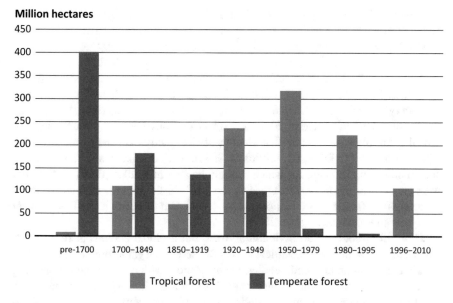

Fig. 1.5 Relative rate of deforestation by year and climate zone. Vertical numbers are in millions of hectares. Source: State of the World's Forests (2012)

Roughly half of the world's cut timber is burned for fuel, including 80% of Africa's. The reasons include a growing export of tropical timber, mainly for the pulp and paper needs of industrialized nations. Per person, developing countries use an average of 6 kg (13 lbs) of paper in a year. The US uses 257 kg (566 lbs) per capita (Zakharov 2014). Furthermore, poor countries are forced to take such measures to improve the balance of trade and reduce debt. As French President Francois Mitterrand said in 1991 at the opening of World Forestry Congress X in Paris, it's hard to criticize the people of tropical regions for allowing the destruction of forests when they must do so to live.

But man's economic activity not only damages the Earth's flora and fauna. It also harms the **soil**—that universal fundament on which all territorial life is based. Plowing up the land and compacting it under agricultural vehicles leads to its degradation, and, without proper soil management, to complete destruction. The cultivation of the virgin soil of Kazakhstan testifies as one example of the irretrievable harm that can be done. By the end of the 50s, after mere decades of cultivation, the country faced horrible ecological consequences such as widespread soil degradation, wind and water erosion, and dust storms. Around the world, 6–7 million ha (15–17 mln ac) of agricultural land are lost each year due to erosion, secondary salinization and other anthropogenic causes. The loss of *humus*, the fertile layer of topsoil, increases constantly.

In all of human history prior to the Industrial Revolution, humus loss added up to roughly 25 million metric tons (27.5 mln standard tons), while in recent centuries—300 (330) million. Over the last 50 years, however, up to 760 (837) million tons of humus have vanished each year (Rosanov et al. 1990) Soil loss, furthermore, is practically irreversible. The recovery of 2.5 cm (1 in.) of topsoil requires 300–1000 years. For 18 cm (7 in.), it takes 2 to 7 thousand years. As a result, according to estimates by the World Resources Institute, the rate of soil degradation exceeds regeneration by anywhere from 16 to 300 times, depending on region (Meadows et al. 2006).

Add to that the area of agricultural land eaten up by transport infrastructure and construction each year. Global statistics concerning this factor do not exist, but there are plenty of localized examples. The Indonesian capital of Jakarta swallows up surrounding land at a pace of 20 thousand ha (49 thousand ac) a year. Vietnamese urbanization likewise uses up that amount of rice paddy over the same period. In China from 1987 to 1992, 6.5 million ha (16 mln ac) of tillage went towards new construction, in exchange for which 3.8 mln ha (9.4 mln ac) of forest and pasture were cleared for the plow. Each year in the US, 170 thousand hectares (420 thousand acres) of farmland are reallocated for roads. And these are just a few of many such examples (Meadows et al. 2006).

This soil is not only an agricultural asset, but a global ecological resource. It plays a vital role in biogeochemical cycles. It serves as a gathering point for water, a veritable ocean on dry land, feeding the plant biota with moisture and supporting the continental water cycle. It also plays host to a plethora of soil organisms. A square meter of topsoil 30 cm (just under a foot) thick contains over a trillion microorganisms and spores. These bacteria, fungi and invertebrates provide for the circulation of decaying organic matter, those biogenic elements (also called nutrients) that the biosphere has limited access to. Normally, the nutrient cycle of a soil ecosystem functions as a **closed loop**, supporting the synthesis and decay of organic matter with a high degree of accuracy which ensures its stability over the course of millennia. In pulling up nutrients along with the harvest, man is constantly exhausting the soil and is forced to support fertility artificially, providing nutrients in the form of fertilizer. If we consider that 11% of land is used for agriculture, and 28% of that (1.4 billion ha/3.46 bln ac) goes under the plow each year, and that disruption to the closed loop biogeochemical cycle on such land goes upwards of 10%, then you must realize the scale of destruction to the biospheric balance that modern agriculture represents.

One of the consequences of ecosystem destruction is the process of **desertification** which now represents a grave problem the world over. Arid, or dry, lands make up 35% of the world's landmass, and over one billion people live on them. Their fates directly depend on the condition of frail and delicate ecosystems, which is what makes desertification such a threat (Fig. 1.6).

This process usually develops as the result of joint actions by man and nature. Elimination of sparse vegetation by overgrazing livestock, chopping down trees and shrubs, and tilling land poorly suited to agriculture all violate the already unstable natural balance. This leads to the degradation of native ecosystems, the drying out and salinization of soil, and then wind erosion. Any ill-considered business in this zone could have disastrous consequences for the natural environment and local populations.

A remarkable example of this is the Aral Sea ecological catastrophe. After many years of using the entire flow of the rivers Amu Darya and Syr Darya for cotton growing, the Aral Sea dropped 20 m from its 1960 level. The salinity in the lakes

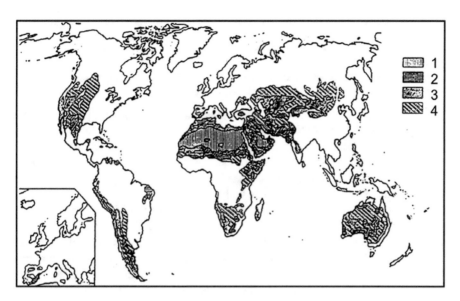

Fig. 1.6 Territories subject to desertification. (1) Deserts; (2) Very High Risk; (3) High Risk; (4) Moderate risk. Source: Maksakovsky (2008) Book 1

that replaced the Aral increased by 5–8 times. The spread of salt by wind and dwindling stocks of groundwater caused a rapid deterioration of ecological wellbeing, salinization and soil degradation over a huge area inhabited by 30 million people.

Another consequence of the misuse of natural resources in these regions has been drought, which has brought disaster to the poorest developing countries of Asia and Africa. These countries, already subject to unfriendly natural forces, are home to 90% of arid zones' residents, half of whom live on the edge of hunger and penury. When drought and famine struck the Sahel, south of the Sahara, in the 1970s and East Africa in the 80s and 90s, hundreds of thousands died. The drought of 2011 dealt a particularly heavy blow to the region (Horn of Africa Drought Crisis..., OCHA 2011). However, the need to feed a large and ever-growing population forces local farmers to abandon developed fields and search out new ones, even though with current agricultural methods the result will likely be the same.

Because of desertification, the world loses about 6 mln ha (14.8 mln ac) of cultivated land yearly. As a rule, these losses are irreversible. UN experts estimate that the desert could claim nearly one third of tillage by the end of the century. By 2025, at the current rate of soil degradation, the continent of Africa will be able to feed a mere 25% of its population (ForexAW.com 2013). These facts have roused the UN to take the initiative. 191 member-states signed the Convention to Combat Desertification in 1996.

Water pollution has already taken on a global scale, and that's just fresh water sources. It has also spread over a large part of closed and semi-closed seas, such as the Caspian, the Baltic, the Sea of Azov and the North Sea. As American ecologist Aldo Leopold wrote in 1941:

"Mechanized man, having rebuilt the landscape, is not rebuilding the waters. The sober citizen who would never submit his watch or his motor to amateur tampering freely submits his lakes to draining, filling, dredging, pollution, stabilization, mosquito control, algae control, swimmer's itch control, and the planting of any fish able to swim. So also for rivers. We constrict them with levees and dams, and then flush them with dredging, channelizations, and the floods and silt of bad farming" (Leopold 1941, p. 17).

Here we must keep in mind that rivers, lakes and the World Ocean mark the final resting place for pollutants that have circulated through city, air and land. Fertilizers and pesticides wash in from farmers' fields. Industrial waste and household waste ends up here. Finally, atmospheric pollutants settle on the surface, deposited by meltwater and rain. So don't be surprised if you find nearly all of Mendeleyev's table in some particularly unfortunate body of water.

Sadly, this applies to many of the arteries of economically developed Europe, despite enormous sums dedicated to their purification. The Elbe, the Oder, the Dnieper, the Southern Bug and the Guadalquivir are all rivers that belong to the category "highly polluted." Pesticides and assorted dangerous organic compounds have accumulated to dangerous levels in them. Concentrations of certain metals such as lead, zinc, chromium and others in the Elbe, for example, are 3–16 times higher than ambient levels (Europe's Environment 1995). The high demand for water further complicates the situation. In some countries, such as Belgium, water processing uses 70% of renewable water resources.

Since 1940 the process of *anthropogenic eutrophication*—the explosive proliferation or "bloom" of blue-green algae[3] due to the accumulation of nutrient elements at the surface—has taken on a massive scope. When an algal bloom occurs, the aerobic bacteria swallows up the oxygen diluted in the water along with dead organic material as they multiply, suffocating the life below and excreting toxins in a wave of death which, furthermore, leads to a sharp decline of water quality.

True, eutrophication also occurs under natural circumstances. But the process in such cases hardly compares with the speed of anthropogenic eutrophication, accelerated by the nitrogen fertilizer that washes off the fields and the phosphorus-rich runoff of urban wastewater. The previous century's hallmark construction of massive dams and reservoirs has deeply compromised the ability of rivers to clean themselves.

Paradoxically, a reservoir can also play a positive role from an ecological point of view. This is particularly apparent at the Volga cascade of hydroelectric stations, which has turned Russia's main water artery into a chain of nearly stagnant reservoirs. These giant basins function largely as cesspools for Volga water. For example, at the Volgograd reservoir, a closed basin 3100 km^2 (1926 mi^2) in area, a bottom sediment 25 cm (9.8 in.) thick had formed by 2007, trapping an enormous mass of harmful and toxic substances (Danilov-Danil'yan and Losev 2006). So, without the reservoirs and at the current catchment area, the Volga would be much dirtier and

---

[3] Cyanobacteria. The two terms will be used interchangeably in the text. These prokaryotic organisms bear a superficial resemblance to algae, which leads to the layman's term despite being unrelated. -*Translator's note.*

could only very generously be called water at all. Today the Volga is considered a "moderately polluted" river and is classified as "polluted" only in some areas. Thanks to the reservoirs, it has higher water quality in its lower reaches than it does mid-course. This paradox illustrates the complexity of human involvement in natural processes whose unpredictability impacts our very survival.

No less a role in water degradation is played by *acidification* and *secondary salinization* of fresh water. The former directly causes what is known as acid rain and results from emissions of oxidized sulfur and nitrogen compounds formed by the combustion of hydrocarbon fuel. When mixed with drops of rain, these molecules react with the water to form sulfuric and nitric acid. This falls on the surface of land and water, often poisoning all life. In any case, withered forests and dead lakes with neither fish nor plankton began appearing in industrial regions of the US, Europe and Japan in the middle of the last century. By the 1970s, they had become a usual occurrence, most often the result of acid rain.

As concerns salinization, well known from the days of ancient Babylon and Assyria, since the twentieth century it has become the scourge of sedentary agriculture. We now use about 1000 (1100) tons of water to produce one metric ton of grain for the worldwide market. If you consider that rice-producing countries use up to 80% of renewable surface and groundwater on agriculture, the result is entirely predictable: a catastrophic lowering of the water table, and salinization of reservoirs thereafter.[4]

In some farming regions of China, the water table is lowering by roughly 1 m per year due to overuse of groundwater. Around Beijing the aquifer has fallen to a depth of 40 m. India is facing similar problems (Maksakovsky 2008, Book 2). Under the twin burdens of booming cities and pollution to surface water, the role of underground water sources has increased dramatically, reaching 50% of overall use in several countries. In many regions of the world, aquifer depletion has already led to serious shortages of fresh water. Meanwhile, demand for this resource is growing faster than population. In order to satisfy the growing demand for food, for example, the share of harvests grown with the aid of irrigation worldwide will have to be 50% higher in 2025 than it was at the end of the last century.

But we already have a deficit of fresh water today equal to the Nile's entire flow of 8 years. According to scientific estimates, 2.7 billion people currently live in river basins subject to severe drought for at least 1 month a year. A particularly difficult "water stress" situation occurs when a period of low water levels coincides with agriculture's seasonal peak demand for water. According to the International Water Management Institute, over a billion people will live in countries with an absolute water scarcity by 2025. The worst effects are in regions of the Middle East, South

---

[4]Along with primary salinization of surface water, it's worth noting secondary, anthropogenic salinization. This arises as the result of irrigation and drainage projects on dry grasslands situated over deep-lying groundwater that rests on saline bedrock. The application of water to the surface opens up previously defunct capillary connections to the aquifer below, drawing up highly-mineralized groundwater. After water has circulated from top to bottom and back, it evaporates, leaving behind a growing layer of salt on the farmland.

Asia, most of Africa and northern China. Even if these regions had highly efficient irrigation systems, they would still not be able to produce enough food on irrigated lands to satisfy their industrial, household and ecological needs. As the authors of the book *Beyond Malthus* noted, "Indeed, the spreading water scarcity may be the most underrated resource issue in the world today" (Brown et al. 1999, p. 37).

Most of Earth's landmass, from the arctic tundra to the burning desert sands, is covered by a continuous membrane of life, **the biota**. This unbroken living quilt resulted from a long process of evolution in which the various species and their communities diverged and adapted to the whole range of geographically and climactically diverse conditions on Earth, as well as their roles within them. This is what we call *biodiversity*, a term well known today even outside academic circles. This is what allows each living being to use the resources available to them within their habitat and *ecological niche*, the "profession" of an organism.

And while the membrane may have ripped in one spot or another at various times during the past due to catastrophic shifts in the planet's crust, volcanic activity or asteroids colliding with the Earth, there have always been forms of life capable of surviving the crisis and filling the breach. This uninterrupted development of life owes itself to biodiversity, the most important factor supporting the biosphere and the efficiency of biogenic processes. By providing the necessary adaptive potential of the biota, biodiversity ensures its survival and future development in a constantly changing environment.

With the beginning of active human economic activity, this priceless evolutionary accomplishment came under threat. The destruction of ecosystems and technological reshaping of the landscape disrupts the ongoing existence of many species and communities, some of which have disappeared from the Earth, and others of which are near extinction. Many species, especially insects and protozoa dwelling under the canopy of tropical forests, die out without even being identified. Even if we limit ourselves to vertebrates, 23 species of fish, two of amphibians, 13 of birds and 83 of mammals have disappeared from the Earth since 1600 (McNeely 1992).

Each extinct species is a final and irreversible loss for the biosphere, and evolution offers no way back. But there are far higher numbers under threat of extinction: 24% of mammal species, 12% of birds and 30% of fish (Species Survival Commission 2001). If this morbid trend continues, it's not hard to imagine what kind of species desert we masters of the planet will have to lord over. Such a biota would also stand little chance of survival after continued material changes to the environment in this degree.

Over the past 20 years, the WWF has developed a program for monitoring global biodiversity on a permanent basis. This "Living Planet Index" allows us to judge the ecological state of the biosphere based on aggregate data for populations of vertebrates in various countries and climate zones. Here in Fig. 1.7 is what the trend looks like of the global Living Planet Index for the period from 1970 to 2008 (1970 is taken as the starting point, so we use it as the baseline value "1").

As you can see from the graph, over the past 40 years the quantitative indicator has declined by almost 30%. That is, the number of wild animals shrank by nearly a third. The situation is especially troubling in the tropical zone, where the index has declined by 60%. Freshwater species of fish declined by 70% (WWF Living Planet

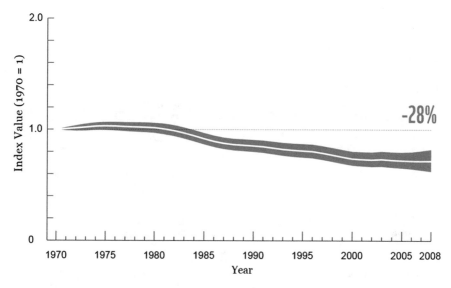

Fig. 1.7 Global Living Planet Index 1970–2008. Source: http://panda.org/downloads/1_lpr_2012_
online_full_size_single_pages_final_120516.pdf

2012). At the same time, the index for the temperate climate zone increased by
30%.[5] That does not mean, however, that ecosystems from that zone are in significantly better shape than those in the tropics. The population index doesn't account
for the tremendous losses suffered in biodiversity prior to 1970. If we could follow
the trend line back several centuries, rather than decades, you would surely see a
drop much like that of the tropical zone today, only extended over a longer period.

Still, growth in the index for the temperate zone tells us about an important
change. People managed to reverse the negative trend by undertaking nature-
friendly programs and events. Since the whaling industry was shut down 40 years
ago, the number of Greenland whales has grown from 1–3 thousand to 10 thousand
head. Wetland and aquatic birds have started recovering in the US. The same is true
for sea birds and migratory birds in the UK (Angliss and Outlaw 2006; Birdlife
International 2008). These welcome tendencies indicate a degree of stewardship
and responsibility towards nature protection by these countries and their neighbors.
Most developing countries lack this emphasis. This is partly due, of course, to limited economic resources, but mainly because there is not the priority placed on ecology which, as a rule, corresponds to the prosperity of the country (Fig. 1.8).

No matter the amount of wrangling there's been over the problem of global
warming, scientists agree that humans are responsible for no less than 50% of the
effect. The lion's share of attention, however, goes to anthropogenic carbon emissions. The role of ecosystems—forests, steppes, wetlands, etc.—that serve as a

---

[5] This data mainly characterizes the state of European and North American populations. Information
concerning wildlife in Central Asia is hard to come by.

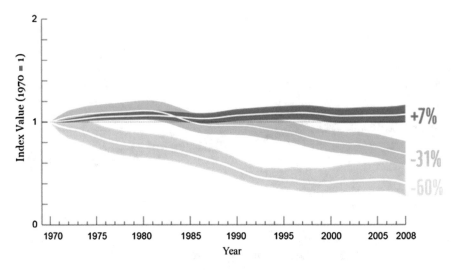

Fig. 1.8 Living Planet Index for three groups of countries according to per-capita income. Source: http://panda.org/downloads/1_lpr_2012_online_full_size_single_pages_final_120516.pdf

natural reservoir for absorbing excess CO_2 and that mankind has efficiently decimated for millennia, unfortunately remains a footnote of popular ecology. You can understand the logic behind that. The chain of cause and effect you typically see when climate change is discussed (increased greenhouse gasses>raised concentrations in the atmosphere>intensification of the greenhouse effect>global warming) is straightforward and quite demonstrative. Most importantly, it provides a clear prescription for the situation: limit the amount of fossil fuels we burn, use alternative sources of energy, encourage energy-saving technology and so forth. But what can be done for ploughed-up steppes and chopped-down forests, which require tens or hundreds of years to recover, assuming we stopped utilizing these lands? What do we do with the deserts that happen to form on the site of forests razed, grown back and razed again until the soil disappears, as is typical of slash-and-burn agriculture? Furthermore, while solving this problem we should remember that the role of the biota in climate change is intricate and complex, involving much more than the absorption of carbon.

Take, for example, the process of active evaporation, or *transpiration*. Clouds form over a forest and water vapor condenses. As it does that, air pressure falls in an atmospheric column and an air mass flows in from the ocean. (For more on this, see Chap. 11.) In this way, violating the ecosystem influences not only the continental water cycle, but the climate system as a whole. The collapse of this mechanism is certain to make itself felt in the most unpredictable ways. We shouldn't only be talking about warming, but of unbalancing the entire climactic machine—a colossal machine so complicated that no computer can model its responses.

Of course, climate systems are highly flexible by their very natures, and their parameters are defined by constant variation around a mean that may itself change over extended periods of time. A totally sustained climate would only be possible on

Table 1.1 Statistics on the largest natural disasters in the second half of the twentieth century

	1950–59	1960–69	1970–79	1980–89	1990–99
Number of natural disasters	20	27	47	63	91
Economic losses in $billions	42.1	75.5	138.4	213.9	654.9

Source: Kondratyev et al. (2005), pp. 57–76

Mars or on the Moon, if only we could apply the idea of a "climate" to them. But in recent decades on Earth, anomalies have become the norm. Cyclones and anticyclones have grown more powerful. They move across larger swathes of land and replace each other less often. Regional irregularity and inconsistency in the climate situation have become typical. Thus, in the US, over the same summer of 1994, scientists noted lowest-ever temperatures on the eastern seaboard while heat records broke on the California coast, reaching 48 °C (118 °F) (Kondratyev et al. 2005).

Add to this picture the anomaly of seasonal shift in the northern hemisphere noted in the middle of the last century. While the timing of change between seasons never varied by more than a day in the previous 350 years, over the past 50 years seasons have come an average of 1.7 days earlier on land than in the first half of the 1900s. Over the ocean, they have begun a day later over the same period. The difference in temperature between seasons has decreased by 2.5 °C. All of these changes are beyond the range of chance variation (Stine et al. 2009).

Most telling of all may be the increasing frequency of natural disasters—floods, droughts, hurricanes, tornadoes, wildfires and others. Each year, upwards of 200 million people suffer their effects, particularly in developing countries. Table 1.1 shows the rate of the most extreme natural disasters in the second half of the twentieth century. As you can see, the number of cataclysms has increased geometrically, claiming tens of thousands of human lives and costing many billions of dollars to clean up. From 1990 to 2015, the yearly number of victims to these catastrophes increased 450%.

This unflagging growth cannot be a coincidence, either. Most climatologists consider this to be the result of climate destabilization connected to human economic activity. According to research conducted by the insurance company Travelers (and insurers take the first monetary losses after tornadoes, hurricanes and floods), raising the surface temperature by a mere 0.9 °C is enough to increase the number of hurricanes on the US coast by a third (van Aalst 2006).

Figure 1.9 uses data from German insurance company "Munich Re" on the increase in natural disasters in the second half of the twentieth century and the accompanying material damages. A decrease in the final years of the twentieth century was paid back with interest in the first years of the twenty-first century. New catastrophes have since created countless victims along with destruction and losses high into the billions.

***

One particular aspect of the global ecological crisis is the stubborn accumulation of waste from human economic activity in the environment, including chemical products with pronounced toxic qualities.

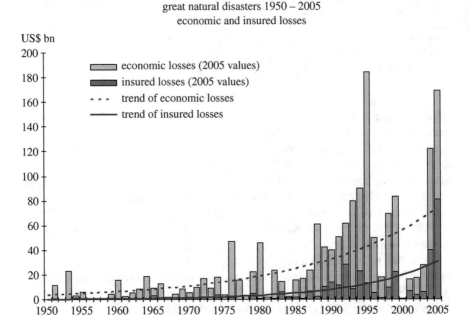

Fig. 1.9 Economic losses from natural disasters. Source: Münich Re

Many people think that pollution itself makes up the greatest threat to modern
civilization (justifiably or not we'll determine later on). Indeed, the amount of waste
has reached cyclopean proportions that beggar the imagination. For each person on
earth, 50 (55) tons of raw materials are yearly called forth from the earth, of which
a mere two metric tons goes into the finished product. Therefore, having undertaken
this enormous labor, humanity gets almost as much back in waste—48 (53) tons,
0.1 tons of it toxic. In developed countries alone, that is 0.5 (0.55) tons of toxic
waste per person (Arsky et al. 1997; Danilov-Danil'yan and Losev 2000).

But the two tons of finished production is also waste, in fact, only transferred to
the future, like a gift for our children and grandchildren. From the ecologist's point
of view, practically everything physically made by man will sooner or later become
a waste product. Just as the Egyptian pyramids and other archaeological sites represent a kind of persistent garbage that allows people to acquaint themselves with
their own history.

Naturally, different forms of waste do not have the same effect on environmental
pollution. In that sense, chemically active substances and their products are beyond
comparison.

Some of them, possessed of high persistency and long half-lives, accumulate in
every medium, including the human body. Others are destroyed in the course of
biological processes, making themselves known only when their intake surpasses
the biochemical ability to destroy them (Odum 1983). Short-lived pollutants (which
dissipate in a matter of weeks) cause regional pollution when they rise into the

atmosphere. If they persist longer than 6 months, the pollution takes on a global character.

Aerosols—tiny, suspended particles between 0.1 and 10 μm in size—are a common contaminant in the atmosphere. They are made up of both solid (dust, ash, soot) and liquid components (sulfur and nitrogen dioxide, ammonia and light hydrocarbons). They absorb toxic high-molecular-weight components and many metals including lead. When introduced to the human respiratory tract, some of them cause irritation or allergic reactions. Others, finding their way into the bloodstream, have a generally toxic effect. Especially dangerous is photochemical smog, a "brown haze" of exhaust and industrial emissions which reacts to solar radiation by producing ethylene, ozone and other unstable molecules.

Hazardous waste and *supertoxicants* represent a special category of contaminants. Industrialized countries produce 90% of these substances, with the USA taking home the gold. However, in recent years, intensive production of hazardous waste has spread to many developing countries, including the rising giants of China, India and Brazil, as well as post-Soviet states such as Russia and Ukraine.

As a rule, countries tend to conceal or keep mum on data about hazardous waste. But, thanks to the efforts of the press, many substances in this group are now household words, including heavy metals and pesticides, as well as related compounds belonging to the chlorohydrocarbon group—dioxins, biphenyls, furans, and others. All of these are very persistent in the environment and, being unknown to the biota, resist chemical or biological breakdown. And so they linger on for decades, invading every sphere and embedding themselves in the food chain that links all earthly species. Dioxins, for example, formed as a byproduct by many technological processes, can be found not only in the atmosphere, soil and water, but also in food, including breast milk from humans and other mammals. As evidence of the truly global proliferation of these pollutants, we witness their discovery even beyond the Arctic Circle, thousands of kilometers from the source emissions. Some of them impact the endocrine, nervous and reproductive systems, for which they are called supertoxicants (The Environment 1993; Colborn et al. 1996; Baranowska et al. 2005).

You probably have some familiarity with the role of pesticides in soil and water pollution. They began their triumphal march with the 1938 discovery of the famed DDT (Dichlorodiphenyltrichloroethane) by Swiss Chemist Paul Muller, who won the Nobel Prize. Mass production began immediately after the Second World War. About 180 brands of pesticide are used in the world today, adding up to 3.2 (3.5) million tons (or just short of 1 1/3 pounds per person) in the 1990s. Developed countries have taken a harder line on pesticides in recent decades, including bans on DDT. Farmers now apply less dangerous forms of pest control. In Third World Countries, however, use of pesticides is not only failing to wind down, but is continuing to increase.

In environmental pathology, pesticides sit at the top of the stress index (followed by heavy metals, transported waste from nuclear plants and toxic waste solids). Generally, between 0.5 and 11 kilos of chemical pesticide are used per hectare (0.44 to 9.8 lbs per acre) of tillage, half of which immediately seeps down into the soil and ground water. In the then-controversial book "Silent Spring" (1962)—one of the first ecological warning sirens—journalist Rachael Carson wrote that the whole human race had come under the influence of chemicals, and no one knew what the long-term consequences might be. Now, 50 years later, the consequences are coming into view.

We've seen, in part, that ecotoxins—whether agricultural herbicides and pesticides (beyond the now-illegal DDT), industry and transport byproducts, such as polychlorinated biphenyls, dioxins, furan etc. or metals such as cadmium, lead and mercury—wind various paths into human bodies, where they wreck untold harm upon the endocrine system, including hormone-associated cancer of the breast and prostate, sperm degeneration, infertility, birth defects and more.

Many of these substances decay slowly and so tend to accumulate in the body. Lead builds up in bone tissue, where in modern humans its concentrations surpass those of our first-millenium ancestors by nearly a thousand times (Khudoley and Mizgiryov 1996). Chlorinated biphenyls build up in fat cells and work their way into breast milk in drops of lipids. As analyses of raw milk samples have shown, even in well-to-do Bavaria, every third sample contains biphenyls at concentrations beyond the acceptable limit (The Environment 1993).

As we've said concerning other issues in this chapter, the "chemicalization of the biosphere" is already a done deal. There are from 100 to 200 thousand different substances floating around the world market, including synthetics and counterfeits. For 80% of them, their effects on living organisms are unknown and unlikely to be completely studied. Passed up the food chain, some of them will accumulate in the upper links (including humans) at concentrations exceeding the initial dose a hundred or a thousand times. So you could very justifiably compare our civilization to a giant animal lab, where the rats are human beings testing upon themselves the effects of some unknown medicine (Coman et al. 2007).

***

But is there any hope in forcing back the raging tide of chemicals that threatens humanity's very existence? And couldn't we use modern technology to somehow overcome the ocean of waste that brought it forth? The first question, we're sorry to say, remains unanswered for now. But as for the widely prevalent illusion that some new technology, even one still in the works, could liberate us from our garbage, we ought to discuss that in more detail.[6]

Let's start with garbage incineration, seeing as it is the most direct and obvious way to eliminate solid waste. It's also tried and true, at over 140 years old. But since the mid-80s many governments in Europe and the Americas have begun winding down this method. Why?

It turns out, first of all, that solid waste simultaneously contains chlorine compounds and transition metals, so the process of incineration produces highly-toxic dioxins. Furthermore, while incineration reduces waste to ashes and slag with a volume ten times lower than before, it produces clouds of gaseous smoke—an average of 6000 m^2 for every metric ton—which contains sulfur dioxide, nitrogen oxides, hydrocarbons and heavy metals in addition to the above-mentioned dioxins. And the whole plume of smoke goes up through the smokestacks into the atmosphere. From there, the air currents carry it for hundreds or thousands of miles.

---

[6] The following section of Chap. 1 was written using materials from K. S. Losev.

Granted, some countries have once again turned to incineration with new technological plant. They presort the garbage and use special filters, along with high temperature incineration technology that prevents the creation of dioxins, benzo(a)pyrene and other burn-off products.

Garbage incineration is but one illustration of the fundamental law of conservation of mass, according to which waste, once produced, can never be eliminated. And clearly it's no coincidence that wildlife produces no garbage as such. The organic byproducts of natural life find their way into a closed food chain, participating one way or another in nutrient cycles. Human waste (aside from that which is universal to the Animal Kingdom, of course) cannot participate in these cycles and thus serves as empty ballast within the biosphere. We can only hide it, bury it, transform it from one phase of matter to another, litter the environment with it, shoot it into space, or, finally, rework it into some new, less-toxic product which, in its turn, will also become waste.

With this in mind, another conventional solution is to create resource-saving technologies or to organize the production system in such a way that one business' waste becomes another's raw material. The famed eco-industrial park of Kalundborg, Denmark, brought such a scheme into existence. Behind the promising facade, however, a portion of unused garbage remains. More importantly, Kalundborg's production is still a form of waste, only put off for another day. The circle, then, has not quite closed. Recycling overall has spread worldwide, with Japan demonstrating the greatest success. Japanese industry reuses about 210 (231) million tons of the country's waste each year, 10% of the total.

Unfortunately, however, all such technologies are expensive and, worse, associated with high usage of energy. All energy production means unavoidable pressure on the environment, and ultimately its deformation and destruction to a degree that negates any positive result.

Japan, again a recognized leader in this field, undertook a structural reform of its economy from the 1970s to the 90s, greatly reducing the role of raw materials and so-called "dirty" industries. Priority was transferred to the information and service industries, high-tech and eco-friendly production built on the principles of recycling, resource conservation and extended product life-cycles. So, what happened? Despite cutting out its own raw materials industry, consumption not only failed to shrink, but even grew. With it grew the mass of accompanying waste. Furthermore, energy usage per capita rose by 15% (Quality of the Environment in Japan 1999). Analogous situations arose in the USA and the countries of Western Europe. Clearly it's no coincidence that the enormous expenditures of the last 40 years on environmental protection and transitioning from "dirty" inefficient economies to "clean" and efficient have not materially reduced per-capita energy usage. On the contrary, in many countries, it just kept going up. Once again, this is a bad sign for the environment.

The widely advertised efforts of various countries to clean local environments have made little difference to the overall global effect. Yes, there have been great successes, such as with the American Great Lakes or the Rhine in Germany, which were in truly horrible shape a half century ago.[7] But, has anyone added up the over-

---

[7] After the second World War, increasing pollution led to a shortage of oxygen in the waters of the Rhine. Levels hit a nadir in 1970, when practically all life in the river was eliminated. By 1980,

all balance of the local clean-ups? How much energy and material was spent upon them, and what were the ecological consequences for the countries they were taken from? Or for the countries the "dirty" industries were taken to?

According to the Law of Communicating Vessels, the ecological gains of one country are often compensated by the losses of others, and so the overall ecological costs, as a rule, surpass the benefits of local cleanups. The WWF's report, "Living Planet 2012," indirectly acknowledges this fact when it says that the ability of rich countries to import resources from poorer ones results in "degrading the biodiversity in those countries while maintaining the remaining biodiversity and ecosystems in their own 'back yard'" (WWF 2012, p. 57). And if the global ecological situation continues to worsen against the background of improvements in a few territories, it resembles nothing so much as "sweeping the problem under the rug" at a planetary level.

And so, it might be time to rethink the second half of the club of Rome's famous slogan: "Think Globally, Act Locally." We need not only to think, but to act globally. Or, at the very least, to review the effectiveness of local actions with a global eye.

As we have seen, in the entire course of its existence Human Civilization has not invented one technology that failed to deform the environment in one way or another. For many long centuries, the biosphere successfully resisted the destructive (business) activity of man. But from the beginning of the twentieth century, the effect of humans upon nature entered a qualitatively different stage; from every side, change toward a decisive end arose as never before witnessed, and it continues to tirelessly accelerate. This means that the compensatory power of the biosphere no longer has the power to resist the influence of civilization, which has grown to ruinous proportions. And this unprecedented ecological crisis has unfolded before our very eyes, in the space of two generations.

References

Angliss, R. P., Outlaw, R. B. (2006). Bowhead whale (Balaena mysticetus): Western Arctic Stock. NOAA's National Marine Fisheries Service, Alaska, National Marine Fisheries Service.

Arsky, Y. M., Danilov-Danil'yan, V. I., Zalikhanov, M. C., Kondratyev, K. Y., Kotlyakov, V. M., & Losyev, K. S. (1997). Ecological problems. In *What is going on? Who is to blame? What is to be done?* Moscow: MNEPU. [in Russian].

Baranowska, I., Barchanska, H., & Pyrsz, A. (2005). Distribution of pesticides and heavy metals in trophic chain. *Chemosphere, 60*, 1590–1589.

---

after major financial investments in purification, things had improved. However, the purification equipment could not deal with toxic heavy metals. This improved only after all of the Rhine countries agreed to harsh laws against environmental pollution. As a result, heavy metals had largely disappeared from the river by 2000, though they remain in silt and riverbeds. A high concentration of chlorine remains, as well as nitrates from farm field runoff. Nonetheless, in 1996 the first salmon was discovered after disappearing 60 years earlier (Weber 2000).

Barnola, J. M., Pimienta, P., Raynaud, D., & Korotkevich, Y. S. (1991). CO2 climate relationship as deduced from Vostok Ice Core: A re-examination based on new measurements and on re-evolution of the air dating. *Tellus, 43B*(2), 83–90.

BirdLife International. (2008). State of the world's birds: indicators for our changing world. Cambridge, UK: BirdLife International. Retrieved from http://datazone.birdlife.org/userfiles/docs/SOWB2008_en.pdf.

Brown, L., Gardner, G., & Halweil, B. (1999). *Beyond Malthus: Nineteen dimensions of the population challenge*. New York: W. W. Norton.

Cannariato, K. G., Kennett, J. P., & Behl, R. J. (1999). Biotic response to late quaternary rapid climate switches in Santa Barbara Basin; ecological and evolutionary implications. *Geology, 27*(1), 63–66.

Colborn, T., Dumanoski, D., & Myers, J. P. (1996). *Our stolen future*. New York: Dutton.

Coman, G., Draghici, C., Chirila, E., & Sica, M. (2007). Pollutants effects on human body: Toxicological approach. In *Chemicals as intentional and accidental global environmental threats*. Dordrecht: Springer.

Danilov-Danil'yan, V. I., & Losev, K. S. (2000). *The ecological challenge and sustainable development*. Moscow: Progress-Traditsia. [in Russian].

Danilov-Danil'yan, V. I., & Losev, K. S. (2006). *Water usage: ecological, econonomic, social and political aspects*. Moscow: Nauka. [in Russian].

Dolnik, V. T. (1992). Are there biological mechanisms for regulating human population numbers? *Priroda, 6*, 3–16. Retrieved from http://vivovoco.astronet.ru/VV/PAPERS/ECCE/VV_EH13W.HTM. [in Russian].

Europe's Environment. (1995). *Statistical compendium for the Dobris assessment*. Eurostat: Luxemburg.

FAO. (2010). *Global Forest Resources Assessment: Key findings*. Rome (Italy): FAO. Retrieved from http://www.fao.org/docrep/013/i1757e/i1757e.pdf.

ForexAW.com. (2013). Retrieved from http://forexaw.com/TERMs/Society/Shocks_and_disasters/Economic_Crisis/l983_%D0%91%D0%B5%D0%B4%D0%BD%D0%BE%D1%81%D1%82%D1%8C_Poverty_%D1%8D%D1%82%D0%BE

Gumilyov, L. N. (2014). An end and a new beginning. Moscow: Ayric-press. Retrieved from http://royallib.com/get/rtf/gumilyov_lev/konets_i_vnov_nachalo_populyarnie_lektsii_po_narodovedeniyu.zip. [in Russian].

Jaworowski, Z. (1997). Another global warming fraud exposed: Ice core data show no carbon dioxide increase. 21st century. *Science and Technology, 10*(1), 42–52.

Joshi, M., Hawkins, E., Sutton, R., Lowe, J., & Frame, D. (2011). Projections of when temperature change will exceed 2°C above pre-industrial levels. *Nature Climate Change*, 407–412.

Kemp, A. C., Horton, B. P., Donnelly, J. P., Mann, M. E., Vermeer, M., & Rahmstorf, S. (2011). Climate related sea-level variations over the past two millennia. *Proceedings of the National Acadamy of Sciences of the United States, 108*(27), 11017–11022.

Khudoley, V. V., & Mizgiryov, I. V. (1996). *Ecologically dangerous factors*. St. Petersburg: Izdatel'stvo "Bank Petrovsky". [in Russian].

Kondratyev, K. Y., & Donchenko, V. K. (1999). *Ecodynamics and geopolitics. Vol. 1: Global problems*. Saint Petersburg. [in Russian].

Kondratyev, K. Y., Krapivin, V. F., & Potapov, I. I. (2005). *Natural disaster statistics. Problems of the environment and natural resourses: General information* (Vol. 5, pp. 55–76). Moscow. [in Russian].

Lashof, D. A., & Ahuja, D. R. (1990). Relative global warming potentials of greenhouse gas emissions. *Nature, 344*, 529–531.

Le Quere, C., et al. (2013). The global carbon budget 1959–2011. *Earth System Science Data, 5*, 165–185. Retrieved from https://spiral.imperial.ac.uk/bitstream/10044/1/41754/3/essd-5-165-2013.pdf

Leopold, A. (1941). Lakes in relation to terrestrial life patterns. In *A symposium on hydrology* (pp. 17–22). Madison: University of Wisconsin Press.

Maksakovsky, V. P. (2008). *A geographical portrait of the world (in two books)*. Moscow: DROFA. [in Russian]. Book 1. Retrieved from http://www.twirpx.com/file/997779/. Book 2. Retrieved from http://www.twirpx.com/file/997899/

McNeely, J. A. (1992). The sinking ark: Pollution and the worldwide loss of biodiversity. *Biodiversity and Conservation, 1*, 2–18.

Meadows, D., Randers., J., & Meadows., D. (2006). *The limits of growth: The 30 year update* (pp. 57–61). London: Earthscan.

National Research Council. (2011). *Climate stabilization targets: Emissions, concentrations, and impacts over. Decades to millennia*. Washington: National Academies Press.

Nobel Prize. (2007). 2007 Nobel Peace Prize Laureates. Retrieved from http://www.nobelprize. org/nobel_prizes/peace/laureates/2007/

Nowinski, N. S., Trumbore, S. E., Schuur, E. A. G., Mack, M. C., & Shaver, G. R. (2007). Nutrient addition prompts rapid destabilization of organic matter in an Arctic tundra ecosystem. *Ecosystems, 2007*. https://doi.org/10.1007/s10021-007-9104-1. Retrieved from http://www. springerlink.com/content/t5650v8x5711187k/

Oak Ridge National Laboratory. (2011). *Carbon dioxide emissions rebound quickly after global financial crisis*. Tennessee, USA.

OCHA. (2011). Horn of Africa Drought Crisis Situation Report No. 5. Retrieved from http://relief-web.int/sites/reliefweb.int/files/resources/Full_report_166.pdf

Odum, E. (1983). *Basic ecology* (p. 518). Philadelphia: Saunders.

Pearce, F. (2005). Climate warning as Siberia melts. *New Scientist*. Aug 11.

Quality of the Environment in Japan. (1999). Tokyo: Institute for Global Environmental Strategies.

Rapp, D. (2008). *Assessing climate change*. Chichester: Springer/Praxis.

Rogelj, J., Hare, W., Lowe, J., Van Vuuren, D. P., Riahi, K., Matthews, B., Hanaoka, T., Jiang, K., & Meinshausen, M. (2011). Emission pathways consistent with a 2_C global temperature limit. *Nature Climate Change, 1*, 413–418.

Rosanov, B. G., Targulian, V., & Orlov, D. S. (1990). Soils. In B. L. Turner et al. (Eds.), *The earth as transformed by human action: Global and regional changes in the biosphere over the past 30 years*. Cambridge: Cambridge University Press.

Siegenthaler, U., Stocker, T. F., Monnin, E., Lüthi, D., Schwander, J., Stauffer, B., Raynaud, D., Barnola, J. M., Fischer, H., Masson-Delmotte, V., & Jouzel, J. (2005). Stable carbon cycle— Climate relationship during the Late Pleistocene. *Science, 310*(5752), 1313–1317.

Species Survival Commission. (2001). *2000 IUCN Red List of threatened species* (p. 2000). Gland, Switzerland: International Union for the Conservation of Nature.

State of the World's forests. (2012). Rome: FAO. Retrieved from http://www.fao.org/3/a-i3010e. pdf

Stine, A. R., Huybers, P., & Fung, I. Y. (2009). Changes in the phase of the annual cycle of surface temperature. *Nature, 457*, 435–441.

The Environment. (1993). *Encyclopedic dictionary and reference*. Moscow: Pangea. [in Russian].

Titlyanova, A. A. (1994). Carbon dioxide and methane emissions into the atmosphere. *Review of Applied and Industrial Mathematics, 6*, 974–978. [in Russian].

van Aalst, M. K. (2006). The impacts of climate change on the risk of natural disasters. *Disasters, 30*(1), 5–18.

Vitousek, P. M. (1994). Beyond global warming: Ecology and global change. *Ecology, 75*(7), 1861–1876.

Weber, U. (2000, June). *The miracle of the Rhine*. UNESCO Courier.

Worldwatch Database. (2000). Retrieved from http://www.worldwatch.org/

WWF Living Planet Report 2012. (2012). In M. Grooten (Ed.). Retrieved from http://www.foot-printnetwork.org/content/images/uploads/LPR_2012.pdf

Zakharov, A. N. (2014). *Development tendencies of real capital in the world and the world economy* (Vol. 4). Russian International Economic Vestnik.

Chapter 2
A Critically Overpopulated Planet

It took roughly half a million years for the species *Homo sapiens* and its hominid forebears to multiply from a few hundred individuals to five or ten million (Kapitsa 1995). In that time, our prehistoric ancestors moved from one foraging site to the next, supplementing their diet with hunting and fishing, for which they acquired the corresponding technologies. Using these technologies, they expanded from the ample forage of subtropical regions into the harsher climate that characterized the Northern Hemisphere in the most recent Ice Age. They were aided in this global expansion by the mastery of fire, the ability to make clothing from animal skins and build shelter from bad weather.

All this allowed our primogenitor to expand its living space, settling all of Eurasia and Australia by 30–35 thousand years ago. Twelve-fifteen thousand years ago, at the end of the last ice age, humans managed to reach the Americas by way of the Behring Strait. They walked over either solid ice or a sea floor exposed by the lowered sea level of the Pleistocene. In this way, the whole planet became the abode of man. The non-genetic inheritance of information—learned experience, technology and skills passed from generation to generation—put humanity in a unique position among the species of the earth.

Mastery of projectile weapons, the bow and arrow, javelin and atlatl, played a vital role in this expansion, allowing humans to hunt large herbivores of the age, especially through the stampede-and-corral method of hunting. Hunters could herd the animals into specially prepared traps where it was easier and safer to make the kill. In the Middle East, you can still see traces of these giant structures from the air, extending several miles into a stone cul-de-sac with narrow runs and pits.

Successful hunting of large mammals was a major cause for the rapid expansion of humanity in what became the *first stage of global population growth*. For the time being, there was no shortage of game. In the southern steppe, in what is now Syria and Jordan, antelope and gazelle roamed in endless herds. The ice age mammoth steppe, besides its eponymous pachyderms, contained ancient bison, muskoxen and wooly rhino. With their advanced weaponry and sophisticated hunting techniques,

© Springer International Publishing AG, part of Springer Nature 2018
V. I. Danilov-Danil'yan, I. E. Reyf, *The Biosphere and Civilization: In the Throes of a Global Crisis*, https://doi.org/10.1007/978-3-319-67193-2_2

ancient humans wiped them out by the thousands. That put humans in the position of monopolist, exercising power over other species.

But ecological monopolies harbor unexpected consequences, including for the monopolistic species itself. The melting of Eurasia and North America's ice cover, beginning ten to twelve thousand years ago, dramatically transformed the landscape and led to the extinction of the native megafauna. This in turn caused the collapse of the Paleolithic hunters' way of life and severe food shortages thereafter. We also cannot discount the likelihood that the ancient hunters themselves contributed to the crisis, exhausting animal stocks through the process known as overkill.

As V. R. Dolnik writes, "Overkill, sooner or later, led them to ecological catastrophe. Narrow specialists, they died out en masse [after the extinction of their prey-*Author's note*], and the survivors reverted to foraging or primitive forms of hunting aside from stampede-and-corral" (Dolnik 1995).

One way or another, anthropologists have clearly ascertained a sharp drop in the number of Stone Age settlements in many regions at the end of glaciation. This has given rise to the hypothesis that hunting tribes were dying out and that the planet's overall human population fell (by some estimates, by up to 90%). This was, however, a very drawn out and gradual process, allowing the hunting tribes to slowly adapt their flexibility, technology and skill to go after smaller mammals, whose herds rushed to occupy a huge area made available by the melting ice sheets.

Human development took a different path in warmer areas with long traditions of foraging culture. There, people invested themselves into gaining maximum independence from the caprices of nature. This happened about 10,000 years ago, at the start of the Neolithic, when man developed agriculture and herding. These fateful innovations entered history as the Neolithic or Agricultural Revolution, a term coined by Australian archaeologist V. Gordon Childe. As Russian mathematician and ecologist Nikita Moiseyev said, "And it truly was a revolution. Humanity not only managed to weather the crisis, but mastering agriculture opened the first page of modern civilization. Ever since, the whole history of the biosphere has gone along a new path, for man decided to create an artificial matter cycle unknown to the pre-human biosphere" (Moiseyev 1998).

According to biologist Nikolai Vavilov's hypothesis, the original geographical center of modern agriculture was the river valleys of the fertile crescent, particularly the headwaters of the Tigris and Euphrates in the Anatolian Plateau (modern Turkey). This is where excavations have revealed the earliest known cultivation of wheat. In this way, ancient farmers provided themselves with food on a completely different basis, moving from mere collection to conscious production. This caused the *second stage* of global population growth.

True, this growth was initially limited to the eastern Mediterranean and valleys such as the Indus and Yellow River where the first civilization took root. But over the following 10,000 years, humans increased their numbers by more than an order of magnitude, from roughly ten million to 200–250 million at the dawn of the Common Era (Maksakovsky 2008).

This increase occurred despite the high child mortality brought about by the rapid spread of disease in early urban centers. This was compensated by the extension of life and an increased birthrate as women lived longer. Furthermore, these patriarchal homesteads no longer viewed children as only "a gift of the gods," but as necessary helpers for their parents. Farming peoples therefore developed an institution of super-fertility (the biblical "be fruitful and multiply" dates to this era) with an average of six to eleven children per woman, two to three of whom would survive to adulthood.

As Dolnik writes, "Unlike us, our ancestors had many brothers and sisters, though not growing up with them, but lying in the grave" (Dolnik 1992). The high birth rate, that is, barely sufficed to cover child mortality, and population growth at the time was no more than 0.05%.

This slow rate of global population growth extended one and a half millennia into our Common Era, impeded by great migrations, grueling wars, blights, famines and epidemic outbreaks that decimated wide regions from time to time. There were centuries, such as the thirteenth, when the population didn't grow at all. The following century, as the pandemic Black Death wiped out a quarter of Europe's inhabitants, it even shrunk. As a result, 500 million people lived on Earth in the year 1500, a mere doubling from the start of the Common Era (Maksakovsky 2008).

The turning point in the demographic trend, first affecting Europe, occurred during the Age of Discovery, when the birth of industry, city growth, medical advancements and improvements to agriculture began to have an impact. This tendency made itself especially clear from the second half of the eighteenth century, when the industrial revolution began to transform European society and created the conditions for rapid population growth.

Along with liberal economic and political systems, the industrial revolution brought with it entirely new technologies for factory production, agriculture, medicine and hygiene, enabling an increase in the length and quality of life as well as a decrease in child mortality. Education also played a major role, not least because it spread concern for sanitary conditions. At the same time, the introduction of previously unknown crops from the Americas dramatically improved the diet of the average Old-World resident. Because of European sea travel and colonial expansion, these developments quickly spread to every continent, becoming a planet-wide phenomenon.

All of this created the conditions for the ongoing third stage of global population growth and our sharp increase in numbers. In the four centuries from 1500 to 1900, the population quadrupled, passing the billion-person mark for the first time in about 1820 (see Fig. 2.1).

Soon, however, the negative side of this process appeared, as the traditional emphasis on high birth rates caused hyperbolic population growth (Cohen 1995; Kapitsa 2008). So, from a population growth rate of no more than 0.05% at the beginning of the common era, by the end of the nineteenth century, it had risen by ten times(!) to 0.5%. Europe began to suffer a growing demographic crisis, resolved in part by emigration to the New World, Siberia and Australia.

A surplus of labor created the basis for a determined economic expansion since it allowed employers to keep pay rates to a minimum. Employers hadn't guessed yet that this was not always in their own best interests (Danilov-Danil'yan 2001). Europe's birth rate peaked in the nineteenth century. At the same time, a bottom-weighted population eased military recruitment and brought an energetic young

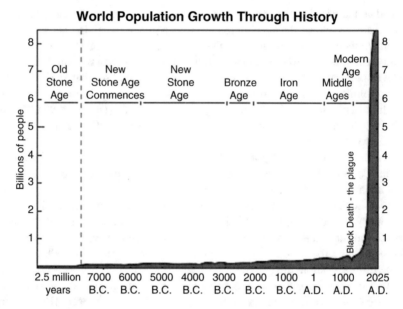

Fig. 2.1 Earth's population growth, beginning at the Stone Age. Source: Penelope ReVelle, Charles ReVelle (1981) Ecology: Issues and Choices for Society. Van Nostrand

generation onto the historical scene. This included a generous helping of those Lev Gumilyov called "Passionaries," charismatic personalities seized with creative energy and demanding urgent social change. Whirlwind demographic processes in many ways set the tone for storms of war and revolution, both in Europe itself and in the colonial possessions, as nations fought to divide other continents.

***

By 1900, the yearly growth rate had reached 0.8% and Earth's population had reached over 1.6 billion. Of that, colonial and dependent nations made up 1070 million, while developed countries (including Russia and Japan) had 560 million (World Resources, 1990–9) The twentieth century brought with it the gradual stabilization of population growth in Europe and North America. A new population strategy— "low birth rate, low death rate, high life expectancy"—appeared as a belated reaction to improved living conditions, reduced infant mortality and achievements in medicine and hygiene. The gradual lowering of the birthrate to one to two children per family continues today in developing countries, leading to a sharp decrease in growth rates for native populations. Practically all growth in those countries now owes itself to immigrants.

But twentieth century demographic trends worked out very differently in "third world" countries. There, too, thanks to higher medical and hygienic standards, water purification and anti-hunger efforts, infant mortality fell and life expectancies rose. In Europe this process extended over two or three centuries, while in developing countries it unfolded in mere decades, leading to destabilization of the demographic situation.

Table 2.1 Earth's population growth rate for the second half of the twentieth century and early twenty-first century

Years	Population growth, %	Years	Population growth, %
1950–1955	1.79	1980–1985	1.71
1955–1960	1.84	1985–1990	1.71
1950–1965	1.98	1990–1995	1.49
1965–1970	2.04	1995–2000	1.35
1970–1975	1.93	2000–2005	1.23
1975–1980	1.72	2005–2010	1.16

Source: World Urbanization Prospects The 2001 Revision, http://www.un.org/esa/population/publications/wup2001/wup2001dh.pdf

As a result, developing countries belatedly adopted a population strategy which Europe had already discarded— "high birth rate, low death rate, high life expectancy." This has caused extreme social tension with periodic explosions. Meanwhile, by the 1960s yearly growth rates had hit 2.5%, a boom larger than anything Europe ever experienced.

The period from 1960 to the 1980s also saw unheard of growth in the overall population of Earth by one and a half times. And while the percentage growth rate has been slowly falling since the 1970s (see Table 2.1), absolute population growth continues to increase, having reached 90 million people a year by the end of the twentieth century. More than ever, this seeming contradiction is explained by the size of the earth's population overall, which almost doubled from 1970 to 1995.

The last decade of the twentieth century proved momentous for global demography. The population strategy of "Lower death, lower birth and lower population growth rate," long initiated in developed countries, began to overtake the world as a whole. This process of changing from the unstable balance of high birth and death rates to low birth and death rates takes the name *demographic transition*, a term that American demographer Frank Notestein coined in 1945.

In Fig. 2.2 we see schematically depicted the demographic transition of both the developed countries of Europe and North America in the nineteenth-twentieth centuries and developing countries in recent decades. From the diagram you can see the distinction between the two: The current demographic transition is going twice as fast as the one in Europe a century ago. The number of people experiencing this arc is 15–20 times higher than those in developed countries.

Here we must comment on two particularities. The first is the reduction of death rate, which occurred twice as fast in "third world" countries than in developed countries. These countries received new medical, sanitary and agricultural technologies ready to go—not only invented, but mastered. The second is the maintenance, despite reduced mortality, of traditional demographic behavior focused on high fertility. Out of 145 million children born each year, 125 million arrive in developing countries (Maksakovsky 2008). This discordance of change—the lowering of mortality and continuing increase in fertility—has led to an unprecedented demographic explosion. Most developing countries turned out not to be ready on a social or psychological level for the introduction of alien innovations that serve as the engine of demographic transition.

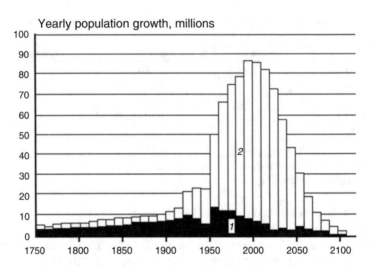

Fig. 2.2 Global demographic transition. From UN data. Yearly population growth rate by decade from 1750 to 2100 (including estimated forecast). Black columns (1) represent developed countries. White columns (2)—developing countries. Source: Kapitsa (2008) A Theoretical Outline of Humanity's Growth. The Demographic Revolution and the Information Society 2008 http://spkurdyumov.narod.ru/kapitsa555.htm [in Russian]

The countries that suffer most from demographic explosion are also the world's most economically and culturally backward. The ceaseless increase of population devours all surplus production, condemning the people, in the attributed words of Jawaharlal Nehru, to "run in place" or often fall even lower than an already beggarly condition. But such is the nature of inertial demographic growth which is, as Dolnik says, "controlled by biological mechanisms, a very complicated population system maintained by custom, tradition and religion. The population demands time, a few generations, to bring birth rates into accordance with death rates" (Dolnik 1992). Only, does humanity have enough time for that?

But the demographic crisis presents problems beyond food and economic development for countries of the "third world." As once occurred in Europe, the population structure here has changed in a historically brief period of time, giving dominance to the younger generation. It was this generation that produced the leaders of anticolonial movements. (Most of these countries were, after all, colonial dependencies until the middle of the last century.) These leaders formed political parties, often of an extremist bent, which soon came under the control of low-to-mid level military officers. These men, in turn, on coming to power, established quasi-military dictatorships, whose outsized ambitions against a backdrop of crushing poverty, social inequality and ethnic strife in no small part brought about the local rebellion and internecine warfare that cut huge swaths through Asia and Africa. It's no coincidence that the lion's share armed conflicts since World War II have broken out in the poorest regions on Earth. Between 1945 and 2000, over 50 million people died in them, 80% civilians (Worldmapper, War Deaths 1945–2000).

The militarization of economics in most developing countries comes as a result of this smoldering violence, accompanied by the steep growth of military budgets. In Africa, just in the 1980s, arms spending more than doubled. These outsized expenditures imposed a terrible burden on already struggling treasuries, denying them means for the few social programs that poor countries are allowed. The poorest per-capita nations, as a rule, spend a greater part of their income on defense than the richest.

We find another inevitable companion of militarization in sovereign debt, a burden on both present and future generations. In the present decade, no less than one fourth of developing countries' foreign debt has gone towards importing weapons. And not the army, but the average citizen, is going to have to pay for it.

Surely it is also not possible to discuss the wave of international terrorism ravaging the world without taking demographics into account. Of course, the lack of opportunity in life, feeling of immiseration and indignity, the aggressive sowing of culturally alien standards are all fertile soil for the next Bin Laden. "These young men," as GEO Magazine put it, "grew up in an atmosphere of desperate rage, which psychologists claim exacerbates narcissistic personality traits to the point of losing the healthy instinct for self-preservation" (Kuklik et al. 2002). Finally, we can't discount the phenomenon of ressentiment, that sense of impotent envy and hatred for "the enemy," in whom the aggrieved sees the source of their misfortune, and which Nietzsche considered the determining characteristic of a slave morality.

From the other side, just as the survival of a given individual in any biological population becomes less important as its numbers go up, so too, clearly, does the individual human life lose value when high birth rates cause overpopulation. A number of researchers (Severtsov 1992) believe that it is possible that our genetic code contains information regarding the ideal population density over a certain area, and that going far beyond that point may negatively influence human psychology. With that in mind, we cannot rule out the unfortunate demographic situation of overcrowded slums in Nigeria or Uganda, Yemen or the Palestinian Territories as a factor in the psychological deformation of suicide–"shahids" and their monumental indifference towards the lives of themselves and others

In this way, decolonization effectively did not bring the promised life improvements to the people of newly-liberated countries, and a chief reason for that was subsequent explosive population growth. That brought on a new age of war and revolution, not unlike what Europe had experienced in previous centuries. So this unprecedented population growth turned out at once to be a stumbling block on the path to economic development and a factor in environmental pressures resulting in social and ecological problems.

***

But let us return to the demographic situation worldwide. At the very turn of the third millennium, humanity passed the six billion mark. In 2011, it passed the seven billion mark. Reaching the first billion people took humanity millions of years, but that number doubled in the next 107 (in 1927). A mere 33 years sufficed to increase by another billion. The next billion-person addition happened 14 years later, with the following billion coming after another 13. Finally, the six and seven billion lines

Table 2.2 Global population growth from the early nineteenth century

Year	Population in billions	Years to increase to next billion
1820	1	All previous human history
1927	2	107
1960	3	33
1974	4	14
1987	5	13
1999	6	12
2011	7	12

Fig. 2.3 Relationship between body mass and populations of mammals, from mice to whales. The shaded zone represents the area of correlation between the average body mass of adult individuals and their approximate population. The arrow represents how far the current numbers of species *Homo Sapiens* have surpassed those corresponding to the rule of natural selection. Source: Akimova and Khaskin (1994)

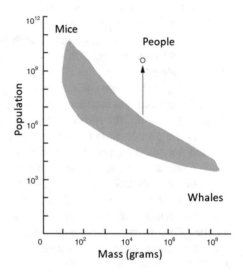

were crossed in 12 years each, in 1999 and 2011 (Table 2.2). It was only in the last 20 years that this feverish growth rate began flagging, falling to 1.3% in 2000. The population is currently increasing by 1.16%, or 79 million people a year in absolute terms. Ninety-five percent of that growth is in the poorest countries on Earth (World Population Prospects 2017) Looking ahead a few decades, the UN predicts a global population of nine billion people by 2050.

Now, if you please, it's about time that we compared these astronomical statistics to what we observe in nature. Humans, after all, for all their technological might, are but one of the species residing on Earth. All of the rules concerning the limits of population growth within a sustainable biosphere apply to us as much as any other creature.

In a balanced environment, each species has its own appropriate population range prescribed by the rule of homeostasis. In part, the numbers depend on the size and mass of organism within a species (Fig. 2.3).

As a large mammal, Homo Sapiens have already gone beyond the biologically typical populations of species with similar body weight by four to five orders of magnitude (Akimova and Khaskin 1994). The biomass of humans and the domesticated animals they keep now makes up 20% of the biomass of all land species. At the beginning of the twentieth century, that number was 1–2% (Warmer et al. 1996). How did this even become possible? Where will it all end?

Biologists know of two reproductive strategies typical of most organisms. The first, the *r-strategy*, is typical of small mammals, for example. Under this strategy, population numbers undergo sharp fluctuations. During the rise, the species exhausts all food supplies, leading to a precipitous drop in the overall number, following a kind of "Boom>Bust>Stabilization" model. By undermining its own means of survival, the population repeatedly passes through the bottleneck, again and again. Among other species, many rodents follow this strategy, such as squirrels and lemmings. We say that this population strategy is high-entropy because it results in a high degree of culling and death, the living material rotting away to mortmass[1] (Krasilov 1992).

The opposite of this is the K-strategy, more typical of large mammals. This boils down to supporting stable numbers and population density through low fecundity, low mortality and long individual life expectancies due to natural defenses (Severtsov 1992). For example, here is how Canadian naturalist Farley Mowat describes the K-strategy-using Arctic Wolves:

"Until they are of breeding age most of the adolescents remain with their parents; but even when they are of age to start a family they are often prevented from doing so by a shortage of homesteads. There is simply not enough hunting territory available to provide the wherewithal for every bitch to raise a litter. Since an overpopulation of wolves above the carrying capacity of the country to maintain would mean a rapid decline in the numbers of prey animals—with consequent starvation for the wolves themselves—they are forced to practice what amounts to birth control through continence. Some adult wolves may have to remain celibate for years before a territory becomes available" (Mowat 1963, p. 180).

It's clear that the K-Strategy fits humans as a species. Nonetheless in various regions and a number of historical periods, we see just the opposite, a shift towards the r-strategy. Explaining this phenomenon poses no difficulty. It all has to do with the artificial expansion of the carrying capacity of new territory by human adoption of more and more advanced technology. As we saw above, one of the first "discoveries," stampede-and-corral hunting, is supposed to have early on played a cruel joke on human populations. By speeding the natural extinction process of ancient predators and megafauna through intensive extirpation, the primordial hunter undermined his own food source, which put the survival of the entire population under threat.

But once people learned to work the land, their food base became decidedly more dependable and grew steadily, sometimes running ahead of population growth,

---

[1] Mortmass is the "waste" of biotic societies, the buildup of dead organic matter. In ancient prehistory, this matter served as the source for the formation of raw hydrocarbons—oil, coal, etc. Coal is thought to have formed from the plant matter of ferns, mosses and rushes in bogs of the Devonian or Carboniferous ages. However, we can observe a tendency toward the reduction of mortmass in the evolutionary process. Unlike these ancient societies, modern biota do not produce this type of waste. Mortmass is instead effectively decomposed by bacteria and fungi.

and other times falling behind. In our own time, the "Green Revolution" of the 1960–80s played an important role in supplying provisions for both developed and developing nations. Governments, with the aid of mostly-Western scientists, introduced high-yield variety crops, modern irrigation methods, streamlining and mechanization, fertilizers and pesticides, all of which sharply increased agricultural production. Thus, in the second half of the twentieth century, global fish and meat production multiplied by five times, soy beans by nine times, and grain by three times, lifting the shadow of hungry death from millions of "third-world" residents. In the period from 1950 to 1984, grain production on Earth grew by 3%, outpacing population growth. Grain consumption per capita increased from 247 to 342 kg (544–754 lb) (Brown et al. 1999).

Supplying all of these mountains of food, however, would have been impossible without a simultaneous increase in energy usage. Hardly any farmers work the land with raw muscle alone any more, and agricultural machinery demands barrel after barrel of combustibles. Agricultural processing, transport and storage are likewise inseparable from energy expenditure. Therefore, to increase food production on Earth by a mere 2%, energy usage needs to increase by at least 5%.

But this is just the tip of the iceberg, only 10% of humanity's energy needs. Because humans, unlike any other living thing on Earth are able to use energy sources other than food. The lion's share, 90% of energy expenditures, goes to specifically human wants—heat and light at home, mechanization of labor, leisure activities and so forth (Rabotnov 2000). And that is why the horsepower of civilization is ever increasing, running far ahead of population growth itself. The population grew from 1.6 billion to six billion or four times over the course of the twentieth century, while energy usage increased roughly ten times over the same period (Vishnevsky 2008). And unlike the demographic boom, the explosion of energy almost entirely occurred in a single century.

This is how, according to data from the International Energy Agency for 2011, the various sources that feed our energy stream break down (Fig. 2.4). Four and a half percent of global energy usage is covered by alternative energy sources—wind, geothermal, solar and biofuel plants—and almost another 16% comes from hydroelectricity. The rest of the nearly 80% is made up of non-renewable sources of energy, including raw hydrocarbons (oil, gas, coal and shale) at about 68% and atomic energy at just short of 12% (International Energy Agency 2013). It is these, the foremost of all energy sources, that serve as our "magic wand," the secret of human power and allows people to overcome the strict species limitations that the biosphere imposes on all living things.

But today this unbridled energy growth has apparently hit a wall. Not only are the vast majority of our energy needs supplied by unrenewable, finite fossil raw materials, but the lengthy efforts to master controlled nuclear fusion have run into a dead end (Yakovlyenko Yakovlenko 1992, 1994). There are even strong doubts about whether it is possible at all. Furthermore, in a world with an expanding fresh water deficit, even fission proves extremely demanding of that resource. Beyond that, expansion of energy demand cannot further follow population growth for reasons unrelated to resource limitations. Energy usage has already approached the

Global Energy Consumption by Source, 2012

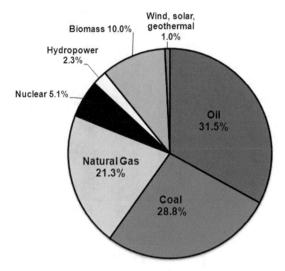

Source: International Energy Agency (IEA 2013)

Fig. 2.4 Global electric production. Source: International Energy Agency (2013)

critical point, beyond which it will irreversibly unbalance the Earth's climate. (See Chap. 14 for more on that.)

***

About 40 years ago, on the threshold of the "ecological age", M. Ibragimbekov's popular play, "The Mesozoic Story," was headlining at Soviet theaters. The play focused on the lives and work of Baku oilmen. The main character, a geologist obsessed with the idea of dredging up oil from deep Mesozoic strata of the Earth, visits the office of his old friend, the all-powerful head of the Baku Oil Trust, in hopes of wheedling money from him for the next exploratory blast.

The cautious and responsible chief shows skepticism at the solicitor's promises. He tries to remind him of the unfortunate results of a previous waterborne explosion—thousands of fish going belly-up. Then the scientist unleashes his most compelling argument, "Fish multiply, oil doesn't."

Well, to borrow the play's wording, fish aren't multiplying nowadays, either. At least they aren't keeping up with population growth. From the diagram (Fig. 2.5), you can clearly see that the global increase in grain harvest totals has been slowing since the mid-1980s. Before that, gains had run ahead of population growth thanks to the green revolution. Now, the quantity of grain harvested per-person is dropping. It dropped by 1% between 1985 and 1995, from a peak of 390 kg (860 lb) per person. At the turn of the millennium, harvests supplied 330 kg (727 lb) per person (Brown et al. 2000). Although overall harvest growth rates have since improved, they have not reached, much less overtaken, the rate of population growth (even as

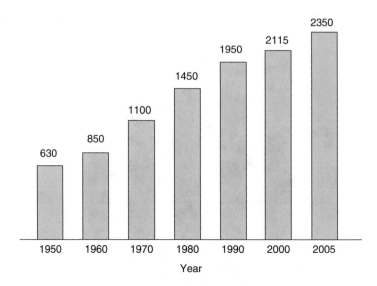

Fig. 2.5 Worldwide grain production, millions of metric tons. Source: Maksakovsky (2008)

that also gradually slows). The momentum of the green revolution, which led to the remarkable gains of the 1960s and 70s, has clearly been exhausted.

But more importantly, the resource of agricultural land is close to exhaustion, along with the fresh water expended on irrigating it. The expansion of tillage began slowing in the second half of the twentieth century, with significant exceptions for the growth of plantations and use of poorer, less suitable land, such as that opened by tropical deforestation.

So, in the period from 1950 to 1981, the acreage used for grain cultivation grew by 25%, from 587 million ha (1451 ac) to 732 million ha (1809 ac). When calculated per person, however, it shrank by 30%, from 0.25 to 0.16 ha (from 0.61 to 0.39 ac). And while overall cultivated land had expanded 15–20% in the half century before the 1990s, population deflated these gains two times over. That is, by the year 2000 per capita tillage fell to only 0.12 ha, half what it was in 1950 (Brown et al. 1999). See Fig. 2.6. If we look several decades ahead, the prognosis tells us that by 2050 there will only be 0.08 ha (0.19 ac) of farmland per person, enough for a small home garden. In some countries, it is likely to be even less, 0.06–0.07 ha (0.14–0.17 ac).

As you can guess, feeding a growing population while reducing per capta acreage is only possible thanks to increasing yields. In 1960, 1 hectare (2.47 acres) produced an average of 2 (2.2) tons of rice, but in 1995—3.6 (3.96) tons. American cornfields yielded a harvest of about 5 (5.5) tons per hectare in 1967, but in 1997—more than 8 (8.8) tons, going as high as 20 (22) tons for some farmers in good years (Meadows et al. 2006).

Irrigation, it must be said, produced truly fantastic results over the course of the twentieth century, in many ways solving the problem of food scarcity. Forty percent

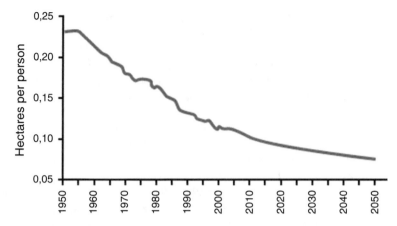

Fig. 2.6 Land used for grain production in hectares per person from 1950 to 2000 with projections up to 2050. Source: Brown et al. (1999)

of the world's harvest at the turn of the millennium was gathered on just 17% of irrigated land (Vishnevsky 2008). However, the long-term success of this mission hinges on a dwindling supply of fresh water. About 55% of yearly fresh water stocks are used around the world today. Of that, 70% goes to irrigation, 20% to industrial needs, 10% to household use (State of World Population 2001). Year by year, competition for water between different sectors of the economy grows more intense. If, for example, a thousand tons of water can be used to either grow one ton of wheat worth $200 or expand industrial production to add $14,000 in value (Brown et al. 1999), the invisible hand of the marketplace will unquestionably guide it to industry. Naturally, under such circumstances agriculture will only survive with the help of government regulation.

By all appearances, this conflict is only going to intensify, and countries living with absolute water scarcity will not be able to maintain 1990-level per capita crop yields as well as satisfy household and industrial demand. They will be forced to import food in ever greater amounts, though this task may be too great for poorer countries. In order to feed the nearly eight billion people on earth in 2025, according to FAO estimates, food production will have to double.

Theoretically, this task is solvable. Several countries on Earth—the U.S., Canada, Argentina, a few European countries, and, of course, Russia—potentially have the ability not only to satisfy their own domestic demand, but to produce significant quantities of food for export. From the other side, there is a group of countries, like Japan, Singapore, and the oil-rich gulf states, which do not possess adequate land and water resources, but have enough money to buy provisions in the necessary amount. But two thirds of the world's people, 3.8 billion of them by the 2016 count, live in countries that simultaneously lack for both food and the money to import it. And those happen to be the countries where the population is growing especially fast.

Obviously, these countries can hardly escape the grip of chronic food scarcity on their own using traditional economic methods. Furthermore, when 70 million new mouths to feed appear each year, it renders meaningless the efforts of the international community to feed the already existent army of hungry people. And even if, some sunny day, the problem of world huger were solved, would it not be a pyrrhic victory? Because its price, as Professor Gretchen Daily of Stanford noted in a copy of *People & the Planet* subtitled "Feeding 8 Billion: Can tomorrow's world feed itself?" could be a destroyed environment (Daily 1995).

***

What will happen to the plant and animal species, whose numbers have fallen critically low in a threat to established ecological balance? The answer is obvious: Stagnation or catastrophic collapse for both the species and the corresponding environmental resources, spelling death for the entire ecosystem.

History has witnessed analogous situations with many species, from bacteria to large mammals. A predator wipes out all of its herbivorous victims. Ungulates trample down all the edible plants in their habitat. But if the population of one species declines under direct influence of prime factors—hunger, environmental degradation, epizootic outbreak (the population r-strategy)—other species are genetically programmed to stabilize population numbers early on, regulating for the *secondary* factors encouraging overpopulation (the K-Strategy). We have already encountered one instance of this regulation—the alteration of mating instincts among wolves during a shortage of hunting territory described by Farley Mowat.

Which category do people belong to? Does biological human nature answer to any kind of secondary factor? First of all, as Dolnik warns, such genetic mechanisms mainly show their influence at a population-wide level, and cannot truly be observed at the individual level. As follows, however often these factors might break through the consciousness of individual people, we can only adequately assess their expression in mass, "statistical" behavior of large social groups (Dolnik 1992).

The same author illustrates the parallel between reactions to secondary factors in animals and peculiarities of human behavior under the conditions of overpopulation. These include, for example, increased aggressiveness of animals during periods of external hardship, or intolerant attitudes towards strangers and outsiders in corresponding situations among humans. They include a lowering fertility rate and ceasing to care for young in overcrowded wildlife populations, and the collapse of the family as an institution in many modern nations. Reactions include the exclusion of a growing number of disabled individuals from the reproductive process. At the final stage, animals lose interest in competition for territory and gather together in a single, floating mass, the "behavioral sink," where reproduction practically stops altogether. Dolnik sees an analogue to this in urbanization, the gathering of people into giant megalopolises which act as a demographic "black hole," noticeably lowering birth rates by the second generation.

Many disagree with this comparison, claiming that in the intervening millennia of socialized existence humans may have lost the corresponding genetic program. But the fact remains: Demographic growth has indeed begun to slow. As Professor A. I. Antonov, Chair for Sociology of the Family at Moscow State University, writes, the process is underway. The system of social norms dictating a high birth rate has collapsed, with mechanisms including later marriages, a more rational approach to sex, the "Contraceptive revolution," removal of limits to terminating pregnancy, premarital relations and divorce, a disconnect between reproduction and sex and so on (The Population of Russia: Near a Dangerous Threshold? 2002).

It must be granted, however, that so far we observe a transition to the K-strategy only in relatively wealthy countries, or even some, such as Russia, where a sharp decline in population could lead to dire consequences. It is not occurring in those places suffering demographic explosion. If the current trend holds, we will have to wait until the twenty-second century for population growth to stop, by which time, according to demographers, there will be ten billion people living on Earth.

One hesitates to consider such a prospect, seeing as the biosphere may not be able to withstand anthropogenic pressure of that scale. People themselves, by that point, will probably run up against the unflinching dictates of primary factors of mortality, control over which had stood as the central accomplishment of humanistic modernity.

Nonetheless, as long as we have the example of developed countries which have almost painlessly completed the transition from "high death-high birth" to "low death-low birth," there remains a shred of hope. At its heart is the greatest social innovation that humanity owes to Western Civilization, the social, economic, and scientific and medical mechanisms to lower mortality. That, in essence, is as much an irreplaceable element of globalization as high technology, modern education or the Internet.

But, as Russian demographer Professor A. G. Vishnevsky notes, "While eagerly following Western experience in the fight against mortality, developing societies, to their misfortune, cannot just as quickly adopt new social mechanisms to limit birth. This is undoubtedly only a delay in an inevitable historical movement. One way or another, having lived through this ruinous period of rejecting 'Western' forms of demographic behavior, they will, in the end, follow—and they are already following—the beaten path of the West" (The Population of Russia: Near a Dangerous Threshold? 2002).

By the way, blindly copying that path is also no guarantee to solve the problem. Mahatma Ghandi, Father of the Indian Nation, understood that many decades ago. One story has it that journalists once asked him if his country, after independence, would reach the same level of prosperity as Great Britain. He answered that on the path to that prosperity Britain had looted half the world. How many planets would India have to loot in order to stand on par with its former colonizer?

The fact is, before transitioning to the new population strategy, developed nations first managed to destroy 9/10 of their own ecosystems, creating powerful environmental destabilization zones in the northern hemisphere. They then went on to annihilate natural reserves thousands of miles from home, turning the rest of the world into a source of raw materials. So, much of the Western path is closed to potential followers. In order to respond to the ecological challenge, developing countries must clearly search out their own path, similar to that of developed countries in some ways, different in others. At this point, there are still too many questions left unanswered. The only obvious thing is that this problem concerns not just developing countries, but humanity as a whole. And there is not much time left to solve it.

References

Akimova, T. A., & Khaskin, B. B. (1994). *Fundamentals of ecological development*. Moscow: Izdatelstvo Rossiyskoy ekonomicheskoy akamedii. [in Russian].

Brown, L., Renner, M., & Halweil, B. (1999). *Vital signs 1999: The environmental trends that are shaping our future*. New York, London: Worldwatch Institute.

Brown, L., et al. (Eds.). (2000). *State of the world 2000*. New York, London: W.W. Norton.

Cohen, J. E. (1995). Population growth and Earth's human carrying capacity. *Science, 269*(5222), 341–346.

Daily, G. (1995). Foreclosing the future. *People and the Planet, 4*, 18–19.

Danilov-Danil'yan, V. I. (2001). Sticker Shock (Surprisy pribavochnoy stoimosti). In V. I. Danilov-Danil'yan (Ed.), *A race to the market: Ten years later*. Moscow: MNEPU. 232 p. [in Russian].

Dolnik, V. R. (1992). Are there biological mechanisms for regulating human population numbers? *Priroda, 6*, 3–16. [in Russian].

Dolnik, V. R. (1995). A right to Land (Article 2). *Znanie—sila* (5–6). Retrieved from http://www.hse.ru/data/2009/10/29/1228458221/Dolnik._Statji.doc. [in Russian].

International Energy Agency. (2013). 2013 Key World Energy Statistics. Paris.

Kapitsa, S. P. (1995). A model of the planet's population growth. *Advancements in the Physical Sciences, 3*, 111–128. [in Russian].

Kapitsa, S. P. (2008). A theoretical outline of humanity's growth. *The Demographic Revolution and the Information Society, 2008*. Retrieved from http://spkurdyumov.narod.ru/kapitsa555.htm. [in Russian].

Krasilov, V. A. (1992). *Environmental protection: principles, problems, priorities*. Moscow: Institut okhrany prirody i zapoved. dela. [in Russian].

Kuklik, K., Luschak, X., & Roiter, K. (2002). The path to heaven is paved with the bodies of infidels. *Geo, 9*, 138–143. [in Russian].

Maksakovsky, V. P. (2008). A geographical portrait of the world (in two books). Moscow: DROFA. [in Russian]. Book 1. Retrieved from http://www.twirpx.com/file/997779/. Book 2. Retrieved from http://www.twirpx.com/file/997899/

Meadows, D., Randers, J., & Meadows, D. (2006). *The limits of growth: The 30 year update* (pp. 57–61). London: Earthscan.

Moiseyev, N. N. (1998). Once again on the problem of coevolution. *Ekologia i zhizn, 2*. [in Russian].

Mowat, F. M. (1963). *Never cry wolf*. New York: Back Bay Books/Little Brown.

Rabotnov, N. (2000). Sorokovka. *Znamya, 2000*(7), 155–174. [in Russian].

Severtsov, A. S. (1992). The dynamics of human population numbers from the standpoint of animal population ecology. Biulleten Moskovskogo obschestva ispytateley prirody. *Otdel biologii, 6*, 3–17. [in Russian].

State of World Population. (2001). Footprints and milestones: Population and environmental change UNFPA, 2001.

The Population of Russia: Near a Dangerous Threshold?. (2002). *Znamya, 5*, 180–199. [in Russian]. Retrieved from http://magazines.russ.ru/znamia/2002/5/nar.html

Vishnevsky, A. (2008). Russia at the Global Demographic Transition. Seminar "Ekonomicheskaya politika v usloviyakh perekhodnogo perioda" [in Russian]. Retrieved from http://www.hse.ru/data/370/900/1235/seminar_27.02.2008.pdf

Warmer, S., Feinstein, M., Coppinger, R., & Clemens, E. (1996). Global population growth and the demise of nature. *Environmental Values, 1996*(5), 285–301.

World Population Prospects. 2017. Retrieved from https://esa.un.org/unpd/wpp/DataQuery/

Yakovlenko, S. I. (1992). Nuclear fusion electric power—an Eternal Engine? *Znanie—sila*, 11–21. [in Russian].

Yakovlenko, S. I. (1994). The problem of energy quality. *Voprosy filosofii*, 95–103. [in Russian].

Chapter 3
The Ecological Footprint of Modern Man

One time in the late 1960s or early '70s, the notable biologist Nikolai Timofeyev-Resovsky made a visit to mathematician Nikita Moiseyev, Assistant Director of the Computing Center at the Soviet Academy of Sciences. He asked him to take the EVM (Electro-Computing Machine—PCs did not exist in those days) and work out how many people could fit into the mineral cycle at the current level of technological development. This was no whimsical notion, and, what's more, it had an agenda behind it. At the time, Moiseyev had taken an interest in the possibility of describing the biosphere quantitatively, as well as the problem of coevolutionary development of society and the biosphere together as systematically connected elements. Timofeyev-Resovsky, for his part, was thinking of how to bring computer modeling methods into biology and was feeling out the interest of mathematicians for that purpose. Finally, the mathematician and the biologist had come together.

"I tinkered with the problem for quite a long time, three or four months," Moiseyev recalled. "Then he called me on the phone and asked if I could tell him anything about the question. I told him there was a high degree of uncertainty, so my answer was inexact, but by my count, it worked out to somewhere between two and eight hundred million people. He had a big laugh and said, 'Almost right—500!' without any calculations."

As it turned out, Timofeyev-Resovsky knew the answer beforehand and wanted to see how a professional mathematician would come to it. Moiseyev continued, "The fact is, only 10% of the energy people use is made up of renewable energy, i.e. the energy that participates in the cycle. Everything else comes from the depository of past biospheres or stores of radioactive materials left over from the Earth's formation. That means that in order not to use up the Earth's reserves, which cannot be renewed, in order not to break the natural mineral cycle and live in harmony with nature like all other living things, humanity would need to either restrain its appetites and find a new technological basis for existence or reduce the population of the planet by nine tenths (Moiseiev 2002, p. 236).

Aside from Timofeyev-Resovsky, a number of well-known scholars—such as Club of Rome founder Aurelio Peccei or American Systems Analysts Jay Forrester,

© Springer International Publishing AG, part of Springer Nature 2018
V. I. Danilov-Danil'yan, I. E. Reyf, *The Biosphere and Civilization: In the Throes of a Global Crisis*, https://doi.org/10.1007/978-3-319-67193-2_3

Donella and Dennis Meadows—have seriously thought about this problem for the first time, perhaps, since the days of Thomas Malthus. What are the limits of anthropogenic pressure before the environment noticeably buckles? How many people, at the current levels of consumption, can the Earth withstand without damage to itself? In any case, it's obvious that humanity has surpassed by an order of magnitude the species population that corresponds to biospheric norms and limitations. And that means we are moving incrementally toward the point beyond which the biosphere's coping capacity will give out and humans will be forced to master other planets. Some futurologists are discussing this possibility in full seriousness.

Nonetheless, the idea that Humanity now feels crowded in its own abode despite all technological conveniences lies at the heart of the concept of an "ecological footprint," proposed in 1992 by scientists William Rees of Canada and Mathis Wackernagel of Switzerland. In brief, an ecological footprint measures a person's impact on the living environment in terms of use of the biosphere's resources and its ability, in turn, to produce said resources and absorb the waste from human activity.

Before pondering this formula, let us turn our attention to the key phrase of that definition, which is the crux of understanding it. It does not concern only the various destabilizing aspects of a person's effect on the environment—pollution, degradation, destruction of ecosystems, etc.—but the *extent of impact* as a universal, quantifiable indicator.

To properly judge the "share" contributed by a given industrial plant, housing block or farm to the destruction of the natural environment, we must first measure it against some kind of unified, standard scale, as one would in most of the natural sciences. The concept of an ecological footprint provides this opportunity. This is because here the negative impact of a person on the biosphere is weighed against a certain common denominator and can be expressed in concrete, comparable units. Further on, based on a sample of the chosen indicators, it becomes possible to estimate the cost to our natural environment of the functioning of individual people, towns, cities and even whole nations. This is called an ecological footprint calculator.

Personal footprint calculators began appearing in the past decade, led by the websites of the World Wildlife Fund and the Global Footprint Network, which many organizations around the world have used as a model. As an example, let's list a few of the questions from a Russian online footprint calculator, which anyone who wishes can answer to add up their "personal" ecological footprint and learn the cost to nature of their everyday habits and ecologically thoughtless behaviors. They see the points accumulating right there. About transportation: Do you use public transport, walk or bike? Do you drive a typical light sedan or a large vehicle with four-wheel drive? About water: Do you take a bath every day or once-twice a week; or do you take a daily shower? For everyday waste: When taking out the trash, do you have separate containers for wastepaper, cans and bottles, plastic packaging, etc.?

Once the calculator adds everything up, you get a result like this one: "Your ecological footprint is equal to 3.1 gha. This is more than nature can provide. If everyone lived like you, we would need 1.5 Planet Earths. For one planet to be enough for all of us, each person should use only 1.8 gha of productive land."[1]

What is a "gha," and how many is 3.1 or 1.8? These are units, called global hectares, corresponding to the statistically *aggregated biocapacity of the planet* in terms of its ability to produce renewable resources and absorb the waste of human activity. This latter refers to carbonoxide gasses, the only waste products currently plugged into the calculation of an ecological footprint, the oft-cited carbon footprint. In this way, a global hectare represents a common unit corresponding to one hectare of average productivity for the Earth. Yet it serves as a measure of biocapacity for many different parts of the biosphere and its associated ecological footprint.]

As you might think, this concept is the product of many thoroughgoing calculations based on great piles of statistical data, which bring together each value involved on the basis of its interconnectedness and relative importance (individual weight). The ecological footprint counts the area that humans remove from natural cycles—farmland and pasture, housing and industrial construction, transport infrastructure, dam reservoirs, forests used as a source for fuel and raw material and fishing zones. On the other side of the equation stands the area of forest and wetlands that absorbs those CO_2 emissions not swallowed up by the ocean. We see the result represented in global hectares that correspond to the area of ecosystems necessary to produce natural resources for, and neutralize byproducts of, human activity (Galli et al. 2007). Sometimes, for clarity, this result is expressed as the time needed to reproduce the renewable resources used and absorb the CO_2 emitted by humanity in a single year or in the number of planets similar to Earth that would have the aggregate bio-capacity to support our global ecological footprint (Fig. 3.1).

In broad strokes, that is the method for calculating the ecological footprint for all of humanity. And we use principally the same methodology to calculate the footprint of any separate entity, such as a farm, town, city, etc.

You can see from the diagram how dangerously the global ecological footprint has grown since 1970, when human demands and nature's supply were roughly equal. But while a half-century ago people, for all of their technology, struggled to fit themselves into the Earth's biosphere, today they have outgrown it by half. If this trend continues into the future on its own inertia, by 2050 the human race will double its ecological footprint and require a whole three Earths.

Returning to a more exact quantitative analysis, in 2008 (the last year for which we had reliable statistics) the Earth encompassed about 12.0 billion gha or 1.8 gha per person. The global ecological footprint equaled 18 billion gha or 2.7 gha per person (The Ecological Footprint Atlas 2010).

---

[1] Of course, many of the parameters used to weigh out anthropogenic influence on the environment have been somewhat arbitrarily determined by the calculator designers, which has often served as a point of criticism for the concept of an ecological footprint. Also, they do not count the way many forms of impact reinforce each other, often in non-linear ways, such as when the effect increases faster than the cause. We certainly cannot take literally the number of planets that correspond to the global ecological footprint. This is, most likely, an attempt to express the situation qualitatively. However, any first attempt to assign quantitative value to that which had never been assigned value before will always give rise to such criticism.

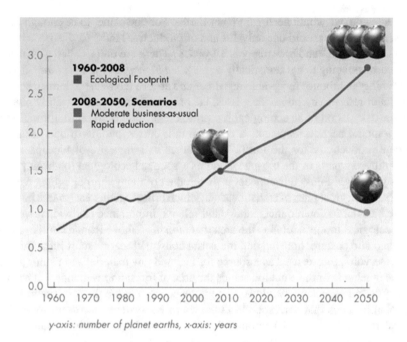

y-axis: number of planet earths, x-axis: years

Fig. 3.1 Global ecological footprint and the number of planets like Earth humanity requires, with the scenarios up to 2050. Source: The Arthur Morgan Institute for Community Solutions: http://www.communitysolution.org/the-100-year-plan/

What is the meaning of this distinction? You'd think an ecological footprint couldn't, or ought not to, exceed the biocapacity of the Earth. Nonetheless, that's what the numbers tell us. The first two characterize the biosphere's potential ability to produce the resources that people extract and utilize the waste that they produce. The latter two express what is actually extracted and utilized. To borrow an analogy from the marketplace, the biosphere's resource production represents the supply of eco-services, and the ecological footprint—their consumption. Thus, in this case, consumption exceeds supply by half, and therefore humanity uses up in one year the amount of resources that the Earth requires a year and half to produce.[2]

"How can this be possible when there is only one Earth?" the authors of the "Living Planet Report 2012" reasonably ask. To explain this paradox, they likewise draw a comparison to the business world. (It's worth noting here that in many ways, this tool for understanding the ecological footprint errs in its anthropocentrism. In reality, nature isn't offering us anything. We'd hazard to guess that the authors use it for the sake of clarity, taking into account the anthropocentrism of the audience they are appealing to.) So, if we pull money out of our bank account faster than

---

[2] The WWF Living Planet Report has been published since 1998, and now comes out every 2 years. It is considered one of the most authoritative sources of information on the Earth's ecological status. The report is developed by scientists from the London Zoological Society and the Global Footprint Network.

interest builds up, our previously accumulated funds will eventually dissipate. Permanent change is happening in the environment in the same way, as we come to the end of supplies of renewable natural resources. We are using them faster than they can regenerate. We can already observe the first indications of this "bankruptcy" today. Symptoms include the intensifying degradation of the environment. They include an unprecedented reduction in biodiversity. Symptoms include increasing concentrations of CO_2 in the atmosphere as a result of ongoing anthropogenic emissions, leading to an increase in average surface temperature, climate change and so on.[3]

But humanity, as you know, is quite varied, and per-person statistics resemble nothing so much as the average depth of a lake. Most rich countries step beyond this measure while poor ones fall well short of it. For example, if everyone lived like the average Indonesian, we'd need two-thirds of the planet's biocapacity. But if everyone consumed at the level of the average U.S. resident, the production of natural resources to be expended would demand four planets like Earth. The average Russian falls somewhere in the middle.

Countries likewise show themselves to be unequal in the maintenance of their ecosystems. In some places, they have been completely destroyed, but fate has shown more kindness to others. Put Brazil and its tropical selva or Russia and its Siberian taiga, tundra bogs and low population density on the scale against China with its powerful technological plant and overpopulation or the U.S. with its hyperconsumption on the other. It seems the first two countries don't fill their "environmental quotas," while the latter live as debtors to more ecologically fortunate countries. This relates to carbon dioxide gas emissions, which are largely swallowed up by donor countries.

Thus we can divide all the world's countries into two groups: 47 ecological donors and 105 recipients.[4] Tables 3.1 and 3.2 showing the top ten of each group are displayed below.

---

[3] Here we'd like to draw attention to conceptual connection between an ecological footprint and the (ecological) carrying capacity of the biosphere (which will be discussed in detail in Chap. 14). In both cases, it is a question of what anthropogenic load the biosphere is in condition to bear while maintaining a fully productive environment. Attempts to estimate the biosphere's carrying capacity have been undertaken based on analysis of energy usage by the insular human technosphere compared with that of biota as a whole (Gorshkov 1980, 1995) or the biomass used by humans (Vitousek et al. 1986). These works, like the concept of a global ecological footprint, are based on hypothetical presuppositions and cannot be regarded as totally well-founded. According to these estimates, anthropogenic impact on the biosphere surpassed its carrying capacity at the turn of the nineteenth-twentieth century, and by the end of the twentieth century, it stood about ten times higher. Then how does the biosphere still exist? Answers to the question are based on a distinction between ecological crisis and ecological disaster. When the anthropogenic load exceeds the biosphere's capacity, it leads to a state of ecological crisis. Up to a certain moment—the point of no return—it maintains the ability to regenerate. Going beyond this point means the irreversible destruction of the biosphere: ecological disaster. The reader should note the similarity to the banking analogy used by the creators of the ecological footprint concept to explain the apparent paradox.

[4] Counting countries with populations of over one million people for which we have reliable statistical data.

Table 3.1 Ecological donor countries 2008 (Ecological capacity exceeds ecological footprint) The order of the countries location corresponds to the size of their ecological capacity reserve

Country	Population (millions)	Ecological capacity (supply)		Ecological footprint (consumption)		Ecological capacity reserve	
		mln ha	ha/ person	mln ha	ha/ person	mln ha	ha/ person
Brazil	190	1708	9.0	552	2.2	+1156	+6.1
Canada	33	492	14.9	231	7.0	+261	+7.9
Argentina	39	296	7.5	103	2.6	+193	+4.9
Russia	142	816	5.7	626	4.4	+190	+1.3
Australia	21	307	14.7	143	6.8	+164	+7.6
Bolivia	10	179	18.8	25	4.9	+154	+15.4
DR Congo	63	173	2.8	47	0.7	+126	+2.0
Colombia	44	177	4.0	83	4.8	+94	+2.1
Peru	29	110	3.9	44	5.4	+66	+2.3
Paraguay	6	69	11.2	20	3.0	+49	+8.2
World	6670	11,895	1.8	17,994	2.7	−6099	−0.9

Source: The Ecological Footprint Atlas 2010 (figures rounded)

Table 3.2 Ecological recipient countries (Ecological footprint exceeds ecological capacity of territory) The order of countries location corresponds to the size of their ecological capacity deficit

Country	Population (millions)	Ecological capacity (supply)		Ecological footprint (consumption)		Ecological capacity deficit	
		mln ha	ha/ person	mln ha	ha/ person	mln ha	ha/ person
China	1336	1307	1.0	2959	2.2	−1652	−1.2
USA	308	1194	3.9	2468	8.0	−1274	−3.7
Japan	127	76	0.6	602	4.8	−532	−4.1
India	1165	594	0.5	1063	0.9	−469	−0.4
Germany	82	158	1.9	418	5.0	−260	−3.1
Italy	58	68	1.1	296	5.0	−228	−3.9
England	61	82	1.3	299	4.9	−217	−4.6
S. Korea	47	16	0.3	233	4.8	−217	−4.6
Spain	44	71	1.6	239	5.4	−168	−3.8
Mexico	107	158	1.5	322	3.0	−164	−1.5
World	6670	11,895	1.8	17,994	2.7	−6099	−0.9

Source: The Ecological Footprint Atlas 2010 (figures rounded)

The three largest "ecological powers" that stand out on the first list are Brazil, Russia and Canada. They determine much of the overall ecological condition on the planet thanks to their enormous expanses of preserved forests. Granted, the ecological footprints of these countries are not very small (Brazil—552 mln gha, Russia—626, Canada—231), but their high ecological capacities cover this with plenty of room to spare (Brazil has 1708 mln gha, Russia—816, Canada—492). But

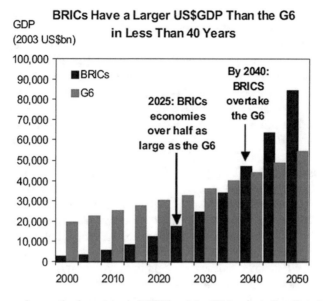

Fig. 3.2 Economic growth of countries in BRICS and the G7 (not including Canada) with forecasts up to 2050. Vertical—aggregate GDP in billions of U.S. dollars. Source: Website the Picky. com   http://www.thepicky.com/investing/ brazil-russia-india-and-china-bric-larger-economy-by-2050/

it is Bolivia that maintains the best ratio of ecological capacity to footprint (surplus ecological capacity or biological sustainability) at +*15.4 gha per capita.*

Three giants lead the second list as well—China, the U.S. and India. These countries stand out not only in terms of natural bounty, but by scale of industrial production and containing over a third of the world's population. Unlike third-place Japan with its limited natural resources, all three countries fall into the top six in terms of ecological capacity (China has 1307 gha, the U.S.—1194, India—1063).

The fact that the BRICS countries (Brazil, Russia, India, China and South Africa) occupy top positions on both lists deserves particular attention. This somewhat artificial grouping plays a growing role in international politics and economics with yearly top-level summits not unlike the more established G-7. These countries stand out in terms of accelerated economic development, occupying a quarter of the world's landmass, having a collective population of nearly three billion, 18% of the world's cumulative GDP and similar metrics. BRICS countries harvest 40% of the world's grain, and two of the countries—Russia and South Africa—hold leading positions in mining and energy markets. China has contended for the title of world's largest economy for several years now, overtaking, for example, the U.S. in its time-honored position as leading auto manufacturer (in cars produced).

It is the future of BRICS that most concerns experts, however, because these countries promise to overtake the G7 in aggregate economic potential in the next 30–40 years, according to estimates (Fig. 3.2). With this in mind, their growing ecological footprints should worry the global community a great deal, along with the maintenance of ecosystems in Brazil and Russia.

CO_2 buildup in the atmosphere and global climate change has already proven a likely cause of anomalous summer heatwaves that struck much of Russia from 2010 to 2014, accompanied by peat and forest fires and a catastrophic flood in the Amur Basin in 2013. Unprecedented droughts struck the Amazon Basin in 2005 and 2010, drying out the tropical rainforests and temporarily turning the region into a net contributor of carbon dioxide gas. Regional carbon emissions into the atmosphere in 2005 added up to 0.8–2.6 gigatons by various estimates, comparable to global emissions from the burning of fossil fuels. The drought of 2010 broke even that record, causing emissions of 1.2–3.4 gigatons of carbon (Lewis et al. 2011).

According to the February 4, 2011 edition of *Science,* "The two recent Amazon droughts demonstrate a mechanism by which remaining intact tropical forests of South America can shift from buffering the increase in atmospheric carbon dioxide to accelerating it... If drought events continue, the era of intact Amazon forests buffering the increase in atmospheric carbon dioxide may have passed" (Lewis et al. 2011). The WWF counted repeated drought in the Amazon Basin and the resultant drying out of rainforests among possible "Points of no return," which might be crossed in coming decades (Lenton et al. 2009). Indeed, due to the rapid decomposition of dead organic material and oxidation of constituent carbon, tropical forests only withhold just over half as much deposited carbon as boreal forests. This is because organic material decays quickly in the wet tropical climate, and the carbon, oxidized to CO_2, returns to the atmosphere.

As far as ecological footprints are concerned, the economic transformation and newfound prosperity of the BRICS states give us plenty to worry about as they increase consumption after the model of more advanced countries. And that is certainly going to tell upon their ecological footprints. For good reason, scientists draw a strong connection between this young century's rapid increase in carbon emissions and the growth of automotive fleets in China, India and Brazil (Oak Ridge National Laboratory 2011).

Any chronicle of environmental footprints would be incomplete if it failed to mention the role of cities. 3.6 billion people, more than half of the world's population, live in cities. Of them, 40% make their homes in cities of over a million souls. The overall number of city dwellers, according to projections, will swell to six billion by 2050 (UNFPA 2007). This, too, should give us cause to worry. Whatever plusses or minuses there may be to urbanization, it brings with it an increase to the ecological footprint. Thus, for example, the average resident of Beijing leaves behind an ecological footprint three times deeper than the Chinese average (Hubacek et al. 2009). This process, dictated by the interests of further developing productive forces, appears consistent and inevitable across the world. As Nikita Moiseyev put it, the growth of megalopolises is not the creation of individuals, but a "natural phenomenon"—a result of society's self-organization.

We can divide countries into three groups based on their extent of urbanization: Those with a city population over 50%, those with 20–50% living in cities and those

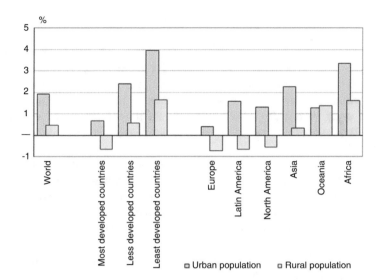

Fig. 3.3 Average yearly rate of growth (or decline) in urban and rural populations by country category and world region, 2005–2010, percent. Source: Demoscope Weekly 2012 http://demoscope.ru/weekly/2012/0507/barom01.php

with less than 20% living in cities. City-dwellers make up an average of 75% of the population in developed countries, but only about 40% in developing nations.

But, while the former have long bid farewell to the days of peak urbanization, developing countries are now experiencing a real urban boom, analogous to the demographic situation overall. This boom reached 2.8% a year at the turn of the millennium, as city population growth shrank to just 0.5% in developing countries (Maksakovsky 2008). As you can see from the diagram (Fig. 3.3), the poorest and least developed countries stood out the most in their rate of urban growth.

In essence, they are following the path of Europe and North America, only more rapidly, with less organization and at greater cost. Furthermore, this game of catch-up is focused primarily in the whirlwind growth of supercities, which suck in the inhabitants of rural backwaters like giant vacuum cleaners without any capacity to provide work or basic social services. The result is a "slum belt" around the city's edge, where migrants from the countryside most often settle into improvised shacks without basic amenities, electrification, running water, or plumbing. Unfortunately, this "slum urbanization" represents this global process in the world's most economically backward regions.

Nonetheless, cities are, in their own way, the engine of civilization, and seeing as most of the world's population has been fated to live in them, it is here that we will decide the destiny of the planet. That includes reducing the size of the collective ecological footprint, which, unfortunately, continues to grow. If the current trend remains in place, growing cities will devour more than half of the century's carbon budget in the next thirty years (WWF 2010).

It is clear, therefore, how much depends on an ecologically literate approach to solving urban problems. These solutions, in the opinion of the "Living Planet Report 2012" authors, should base themselves on "One Planet Principles," an understanding of the limitations of the Earth's resources. These principles include eco-efficient technologies, various forms of energy saving, the use of renewable energy sources such as wind and solar installations, compactness of urban construction to reduce transport costs, prioritizing public over private transportation and encouraging changes to behavior from consumer habits to business models, in order to put environmental interests first.[5]

But how well does the WWF's "City Challenge" fit with the realities of life, particularly those in the former "Third World?" And how do we superimpose the model of some European capital, where the citizenry presorts household waste out of habit, upon ten million-strong Cairo or 18 million-strong Mumbai, where most garbage remains within the city and thousands of residents live directly upon the trash heaps, their only source of income? Or, how will the average Indian or Chinese, having achieved their lifelong dream of buying a car, accept the request to limit driving and climb back on a bike?

It goes without saying that certain advancements in the environmental consciousness of the average European or American in the past 15–20 years and the corresponding governmental conservationist efforts have already borne some fruit. For example, a number of European and North American cities have reduced their relative ecological (particularly carbon) footprint. Thus, New York City's CO_2 emissions have fallen to a 30% lower per capita rate than the U.S. average (Dodman 2009).

But two-thirds of the world's urban population lives in a completely different kind of city and mainly concerns itself with entirely different problems. That is where the pace is set for the growth of the global ecological footprint. And whereas the residents of African cities' expanding outer regions typically lack for electrification and, to a large extent, supply their daily energy needs by the decimation of nearby tropical forests, we can hardly count on any reduction of the ecological footprint there. Just as desperate economic straits give them little chance to break free of the vicious cycle of poverty under their own power and find a way out of this hopeless situation.

As analysis from experts at the World Bank has shown (2014), recognition of the true value of natural resources such as clean air, uncontaminated water, untouched forest or rich soil usually comes together with a certain level of prosperity. Thus, the smoke content in a city's air reaches its highest extent just as a country passes a

---

[5] Here let us introduce a few examples of water conservation for domestic and workplace settings often found in ecological literature. Use different quality water for different purposes (such as rain or drainage water for flushing toilets and watering the lawn). Install a screw-on regulator on your tap to stabilize flow. Fix leaks, which cause the U.S., for example, to lose a fourth of the water flowing through its pipes. Introduce water purification and recycling to industry. Install a water usage meter, which reduces expenditure of water by 30–40% (Hawken et al. 1999).

per-capita income level of roughly $6000 a year. After that, the pollution level falls as a result of appropriate measures being taken, and so on.

Poverty and popular disregard for the natural environment go hand in hand. In that regard, the proactive section of the WWF's program put forth in the "Living Planet Report 2012" better suits the realities of economically developed countries, which are apparently destined to provide the driving force for the rest of humanity. Which of these two incompatible forces will come out on top? There's no answer to that question as yet. We can only repeat the authors of the report when they say, "Implementing such a paradigm shift will be a tremendous challenge, involving uncomfortable decisions and tradeoffs." But the real question is, will it be our own, free, deliberate decision made in time, or a decision forced upon us by the stern dictates of ultimate factors once we belatedly recognize our missed opportunities? It is this question, of which decision to make, that will determine the fate of humanity.

References

Dodman, D. (2009). Urban density and climate change. In J. M. Guzmán, et al., (Ed.), Analytical review of the interaction between urban growth trends and environmental changes (Revised draft: April 2, 2009). New York: United Nations Population Fund (UNFPA).

Galli, A., Kitzes, J., Wermer, P., Wackernagel, M., Niccolucci, V., & Tiezzi, E. (2007). An exploration of the mathematics behind the ecological footprint. *International Journal of Ecodynamics, 2*(4), 250–257.

Gorshkov, V. G. (1980). The Structure of Biospheric Energy Flux. *Botanichesky zhurnal, 65*(11), 1579–1590. [in Russian].

Gorshkov, V. G. (1995). *Physical and Biological Bases for Sustainable Life*. Moscow: VINITI [in Russian].

Hawken, P., Lovins A., Lovins, H. (1999). Natural Capital. New York: Little, Brown, and Co. (Chap. 11).

Hubacek, K., Guan, D., Barrett, J., & Wiedmann, T. (2009). Environmental implications of urbanization and lifestyle change in China: Ecological and water footprints. *Journal of Cleaner Production, 17*, 1241–1248.

Lenton, T., Footitt, A., & Dlugolecki, A. (2009). *Major tipping points in the Earth's climate system and consequences for the insurance sector*. Berlin, Munich: WWF and Allianz.

Lewis, S. L., Brando, P. M., Phillips, O. L., van der Heijden, G. M., & Nepstad, D. (2011). The 2010 Amazon drought. *Science, 331*(6017), 554.

Maksakovsky, V. P. (2008). *A geographical portrait of the world (in two books)*. Moscow: DROFA [in Russian]. Book 1. Retrieved from http://www.twirpx.com/file/997779/. Book 2. Retrieved from http://www.twirpx.com/file/997899/

Moiseiev, N. N. (2002). *How far off is tomorrow?* Moscow: Taydeks Ko. [in Russian].

Oak Ridge National Laboratory. (2011). *Carbon dioxide emissions rebound quickly after global financial crisis*. Tennessee, USA.

UNFPA. (2007). *State of world population 2007: Unleashing the potential of urban growth*. New York: UN Population Fund.

Vitousek, P. M., Ehrlich, P. R., Ehrlich, A. H. E., & Matson, P. A. (1986). Human appropriation of the product of photosynthesis. *Bioscience, 36*(5), 368–375.

WWF. (2010). *Reinventing the city: Three perquisites for greening urban infrastructures*. Gland: WWF International.

Part II
Civilization in Crisis: The Edge of the Abyss (Continued)

Chapter 4
The Social Dimensions of the Crisis

We cannot accurately examine either the ecological or demographic situation on the planet separately from the social crisis, though until recently that was exactly what most people did. They imagined nature as merely the backdrop for the stage on which the social drama played itself out. But today we hardly need to prove the deep connection between the social and natural environments, even if it doesn't always make itself plain. Thus, there are firm grounds for calling the present ecological crisis a socio-ecological one, which we will here attempt to demonstrate.

The various regions of the world each experience this crisis in their own peculiar ways, and the most vital problems of one country may not even be a blip on the radar of others. Take, for example, the most serious social problem of developing countries, *poverty*. We can examine the various degrees of poverty in either relative or absolute terms. Relative poverty is a more relevant question for economically developed countries, which we can define in comparison to the accepted, "normal" standard of living for a given society. The criteria of absolute poverty are primarily connected to physiological factors, the need for the vital resources which preserve the biological life of a person. Based on the criteria of per-person income, the UN and World Bank have established a critical threshold of poverty equal to $1.25 a day in the purchasing power of constant international US dollars. Around the world, 1.2 billion people live in this condition of *extreme poverty*.

At the UN Millennium Summit in New York in 2000, participants set "Millennium Development Goals" for the next 15 years that included battling poverty, hunger, and disease as well as mother and child mortality. The most impressive gains were made in East Asia, where poverty levels fell from 60% in 1990 to 16% in 2005 and continued downward to 12% by 2010. China alone reduced the number of its people living on under $1.25 a day by 600 million people. Worldwide, the proportion of the population fell by half, from 52% in 1981 to 26% (World Bank Group 2011) (Fig. 4.1).

Unfortunately, this process more weakly affected South Asia. Although the proportion of those living in poverty fell from 50 to 30% (1990–2010), the overall number of impoverished remained the same due to population growth. In Sub-Saharan Africa, attempts to relieve poverty went practically unrewarded. Half

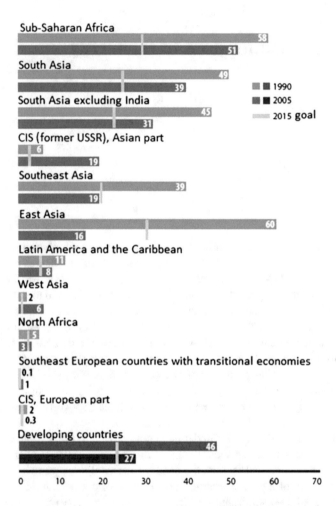

Fig. 4.1 Proportion of people living below the poverty threshold (less than $1.25 per day) in 1990 and 2005 (light and dark bars, respectively). The vertical stripe represents the goal for 2015. Source: Alkire and Santos (2010)

of the population still lives in poverty, and, alone among all regions of the world, the absolute number of the poor rose—from 290 million in 1990 to 414 million in 2010. Here the difficult economic situation suffers further under the weight of ethnic warfare, the HIV/AIDS epidemic and a merry-go-round of coups d'état (Millennium Development Goals… 2013). These countries in particular—The Democratic Republic of the Congo, Zimbabwe, Liberia, Niger, etc.—with average incomes of less than $1000 a year, head up the list of poorest countries in the world. Meanwhile, the distance between the poorest countries and the richest has only increased over the past 40 years.

In 1970, the average income in countries belonging to the richest quartile in global ratings was 23 times higher than those in the poorest quartile. By 2010, the difference had grown to 29 times. Residents of the 13 poorest countries are receiving a lower income on average today than they were in 1970. In Zimbabwe, the poorest of the poor, they make 25% less. As the UN Human Development Report 2010 acknowledges, "The distance between the richest and poorest countries has widened to a gulf." Thus, a person born in Niger, for example, lives an average of 26 years less than one born in Denmark, and consumes 55 times fewer goods (Human Development... 2010).

Poverty tells especially hard upon children, and, as a rule, the damage suffered at a young age is irreversible. We must, therefore, consider the tale of Oliver Twist, composed by jolly old Dickens for lovers of happy endings, a fairy tale having little in common with a decidedly merciless and cynical reality. The statistics brought forth in the Human Development Report 2014 speak for themselves. One in five children in the developing world, where 92% of the world's children are, lives in a condition of absolute poverty according to family income and is especially susceptible to malnutrition. Seven out of a hundred such children never live to age five, and fifty will never have a birth record. Seventeen of them will never go to school, and 30 will suffer stunted growth due to inadequate nutrition. This last factor leads to the fatal outcome in 35% of deadly cases of measles, malaria, inflammatory pneumonia and diarrhea. A lack of plumbing and clean drinking water greatly increases these children's risk of infectious diseases (Human Development... 2014).

The outlook for these children's futures is no less grim. "Lacking basic nutrition, health care and stimulation to promote healthy growth, many poor children enter school unready to learn, and they do poorly in class, repeat grades and are likely to drop out. For children who survive, poverty and undernutrition during preschool years account for a subsequent loss of more than two school grades...When educational attainment is reduced, vulnerabilities are transmitted across generations by limiting children's future learning and employment opportunities" (Human Development... 2014). In this way, by starting a person down a bad road from their first days on Earth, poverty has the ability to self-perpetuate from generation to generation, leaving those caught under its spell little chance of escape.

But how much can we depend upon a purely monetary benchmark like average per-person income? How well does it paint the full picture of a complicated phenomenon like poverty? Poverty and want are multifaceted, expressing themselves in many dimensions. They encompass a lack of opportunity in life, an unfulfilled yearning for knowledge from lack of access to education, a paucity of means to support the health and energy of one's life. Finally, poverty is the absence of conditions to maintain a basic sense of self-respect and human dignity.

All of these considerations prompted scientists to explore new approaches to the problem, resulting in the 2010 proposal of a new integrated benchmark, the *Multidimensional Poverty Index* (MPI). The MPI provides researchers with a more complete "face" of poverty compared with the traditional approach based on income. At present, it is accepted by the majority of countries and most organizations, including the UN, that study issues of poverty and social inequality.

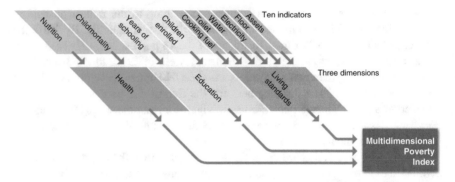

Fig. 4.2 Components of the Multidimensional Poverty Index. Source: Alkire and Santos (2010)

Drawing on three basic data sets concerning health, education and standards of living, the MPI shows how many people at the family or household level (the scale of poverty) are experiencing various forms of want or deprivation, and how many of these forms they are suffering from at once (its depth and intensity). The MPI is used to compare the weight of poverty upon families in various situations, such as one where a 5-year-old child died and the older brothers and sisters don't go to school, or another living in a house without plumbing, water or electricity with an earthen floor and dirty fuel (dung, wood, charcoal) fouling the air inside.

Figure 4.2 shows the three basic measures of the MPI—health, education and standard of living—and ten indicators corresponding to types of deprivation people face. So, the health measurement has two indicators—a lack of food for any member of the household, and the number of children in the family who have died. The education measurement ascertains the number of children who fail to attend school, as well as the number of adults without a fifth grade education. Finally, the standard of living measurement contains six indicators which count earthen floors, the use of dirty fuel for cooking, lack of electrification, lack of plumbing, lack of access to clean drinking water as well as lack of a car or truck and at least one of the following—bicycle, motorcycle, radio, refrigerator, telephone or television.

For a household to be considered multi-dimensionally poor, it should suffer several deprivations at once—at least three of the ten. Scientists then calculate the MPI by multiplying the impoverished share of the population (*H*) by the depth or intensity indicator of Poverty (*A*), giving them the median number of deprivations people suffer (Human…2010).

As we have already said, income data often fails to paint the full picture of poverty and does not contain information about the health and education of the corresponding population groups. The income of paupers does not always convert to education and health, which depend on local conditions not only at the national level, but also at the state or provincial level. Some poor countries, such as Tanzania, Uzbekistan and Sri Lanka, provide medical care and education free or at a nominal price. In others, like Niger or Ethiopia, such services often prove inaccessible even to the gainfully employed. That is how the MPI supplements financial indicators of

Fig. 4.3 Distribution of the multi-dimensionally poor population in developing countries by region. Sample includes 92% of the populations of 98 countries. Source: http://hdr.undp.org/sites/default/files/reports/270/hdr_2010_en_complete_reprint.pdf

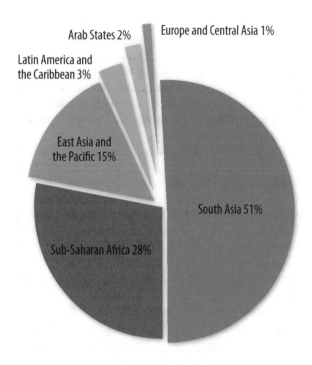

poverty and enables governments to form more effective policies in this area. And while we can observe a correlation between assessments of poverty based on the MPI and the $1.25 per day indicator, these two assessments noticeably diverge at both national and worldwide levels. Thus, according to research, from 2001 to 2010 about 1.7 billion people (almost one-fourth of the global population) lived in a condition of multi-dimension poverty, while only 1.3 billion lived on less than $1.25 a day. Both of these numbers come from 109 developing countries with a total population of 5.5 billion. As we can see, this represents a substantial distinction (UN Human Development Index for 2011).

Sub-Saharan African countries suffer the highest percentages of multidimensional poverty. Worst off is Niger, where the number has reached the monstrous proportions of 92%. Not far behind follow Ethiopia at 89% and Mali at 87%, then another twenty-three of the poorest African states. More than half of families have survived the death of a child, and roughly as many lack even a rudimentary education. In absolute numbers, however, the largest populations of the multi-dimensionally impoverished live in South Asia—Pakistan, Bangladesh and India (See diagram, Fig. 4.3). In just eight of India's 28 states, there live more multi-dimensionally poor people (421 million) than in all of those 26 poorest African countries combined (410 million). And this is despite obvious progress in India's 30-year war on poverty, which increased spending on social services and reduced poverty from 50% in 1983 to 32.7% in 2010 (Human…2010, 2013).

But, it must be added, the most terrible companion of poverty is *famine* which periodically strikes enormous swathes of populations in poorly-developed countries.

Indeed, the threat of famine has hung over the human race like the sword of Damocles since biblical times, pushing whole tribes and nations to the edge of existence. Whole cities and towns died out in the face of drought, disaster and poor harvests, often exacerbated by social upheaval and war. The 1921–22 famine in the Volga Region that killed five million people, a consequence of the Russian Civil War, comes to mind.

Famine periodically came to Europe's doorstep right up to the end of the nineteenth century. The potato blight and subsequent hunger in Ireland from 1845 to 1849 is the clearest example, when a pathogenic fungus struck the staple crop of the Irish poor. Up to 1.5 million people died, and as many emigrated, many to America, directly resulting in a loss of roughly a quarter of the population. Only in the twentieth century, as new methods of agronomy and selection brought higher-yielding crops to the fields, did the threat of famine retreat from Europe and North America for good. Today, countries inhabited by a mere 18% of the Earth's population provide ¾ of the world's food production, with the largest exporters being the USA, Canada, Australia, France and Argentina.

But, you see a completely different picture in developing countries with their rapidly expanding populations, in most of which harvests cannot keep up. Granted, the Green Revolution of the post-war decades significantly alleviated the problem with its high-tech approaches to cultivation. In the 1950s and '60s, grain production even outpaced population growth. In the long term, however, since the 1980s, the increase of grain production has slowed by about 1% per year, which has primarily affected economically disadvantaged countries where the demographic boom continued but suitable farmland melted away like an overused stick of chalk. And while the number of chronically malnourished has decreased, from 920 million in the early '70s to 850 million today, it remains very high. As nearly all those suffering from a shortage of food are concentrated at equatorial latitudes, we might speak of a famine and malnutrition belt, encircling the globe on either side of 0° and including Central America, the Caribbean nations, South and Southeast Asia and all of tropical Africa. In this last region, the poorest on earth, roughly 200 million people suffer from malnutrition, with numbers reaching 30–40% of the population in countries like Chad, Somalia, Uganda, Mozambique, Ethiopia and Zambia (Maksakovsky 2008, Book 1).

One in eight of the world's inhabitants suffer from chronic malnutrition today, and nine million people a year (25,000/day) die from resultant complications. The war against hunger, therefore, is one of the UN's top priorities. With this in mind, the UN Millennium Development goals, approved by member states in 2000, set the goal of cutting the 1990 number of hungry in half by 2015, along with the number of poor earning under $1 a day (Millenium…report for 2013). A number of gains have certainly been made. Since 1990–92, the share of malnourished in developing countries overall decreased from 23 to 15%. In Southeast Asia, it fell from 30 to 11%. In East Asia, including China, it went from 21 to 12%, so we can consider, at least in some places, this millennium goal reached (See Table 4.1).

Table 4.1 Share of people suffering from malnutrition in 1990–92 and 2010–12

Region	Share of people suffering from malnutrition 1990–92, %	Share of people suffering from malnutrition 2010–12, %
Sub-Saharan Africa	32	27
South Asia	27	18
East Asia	21	12
Southeast Asia	30	11
West Asia	7	10
Latin America	15	8
Central Asia	14	7
North Africa	4	3
All developing countries	23	15

Source: Millennium… (2013)

The situation in Latin America was a bit worse. Only in South Asia and Sub-Saharan Africa did efforts fall well short of the mark. But the hope of halving the army of the famished by 2015 clearly did not come to pass. Not only did the global financial crisis of 2008–2010 take a particularly heavy toll on the economies of poor countries, but ongoing factors, such as exhaustion of natural resources, the dwindling of tillable land and shortages of water for irrigation, made themselves felt. We are clearly approaching a ceiling in our use of the World Ocean's renewable resources. FAO specialists have demonstrated the extent to which they have been overdrawn. Nine of the seventeen main fishing areas are on the edge of collapse. Thus, we cannot count on a material increase in the global fish catch either (Maksakovsky 2008).

And so, the ambitious plans to eliminate hunger in underdeveloped countries remain as yet unfulfilled, with many unknowns regarding their execution. Only time will tell whether or not the problem can be solved. *Demoscope Weekly*, the online periodical of the Higher School of Economics' Institute of Demography, expressed it this way, "Furthermore, there can be no certainty that the introduction of efficient agricultural technology, should it become economically feasible, would lead to an end of food shortages. By now we are well aware that applying these technologies often brings about unforeseen consequences and gives rise to new difficulties…In any case, rapid growth of food production under the conditions of general poverty will result in an increased burden on the planet's natural resources, pushing them to the brink of total exhaustion" (*Demoscope* 2002).

Poverty and environmental degradation are inseparably linked to one another. Three-fourths of the world's poor live in rural areas, practicing traditional forms of agriculture. As a rule, they cultivate ill-suited land—dry, steep terraces, infertile soil from destroyed tropical forests and such. Lacking technological and financial means to support soil fertility or to battle salinization and erosion, this leads to the rapid depletion of cultivated land.

The rest of the impoverished, moving into cities, mainly settle on the edge of town among wastelands and garbage dumps. Neither the inhabitants who live on one-two dollars a day, nor the local government, has funds to clean up the urban environment. As a result, solid refuse builds up on the city streets, polluting soil and air, particularly when burned in open fires which produce many toxic byproducts.

As a rule, the housing is poorly furnished, and wood and brush serve as fuel for heat and cooking. Burnt in primitive stoves, this smokes up and befouls the home. The gathering of this wood from nearby forests serves as one of the main causes of deforestation (along with commercial logging). Take the cutting down of forests near Dar-es-Salaam, Tanzania, for example, which has led to the complete elimination of valuable tree species for a radius of 200 km around the city. Spreading at a rate of 9 km/year, this wave of degradation has seriously damaged both the biodiversity and the biological productivity of surviving ecosystems. There are now 70% fewer species in remaining nearby forests than in those further from the city, and the compromised lands absorb 90% less carbon. Thus, with increasing demand for construction lumber and without a cheap alternative for firewood and charcoal, African megalopolises are turning into major centers of environmental degradation (WWF Living Planet, 2010).

Developing countries have a serious problem with providing clean water and plumbing to their populations, the lack of which is closely bound to poverty. At present about 1.2 billion people lack access to quality drinking water in developing countries and a number of former Soviet republics. Twice as many do not have plumbing, which increases the risk of intestinal disease, cholera, dysentery, typhoid and hepatitis. World Health Organization (WHO) specialists estimate that about five million people die as a result of using polluted water each year. As former WHO director Halfdan Mahler noted, "The number of water taps per 1000 people is a better indicator of health than the number of hospital beds." (Danilov-Danil'yan and Losev 2006, p. 100)

But while the number of people without access to safe water is gradually shrinking, the population of those lacking toilets continues to grow, primarily in developing countries. The absence of plumbing leads to worsening fecal contamination of both surface and groundwater and declining drinking water quality, causing the abovementioned five million deaths due to unsanitary conditions (Danilov-Danil'yan and Losev 2006).

Industrial pollution represents another acute problem for developing countries. The arrival of urbanization and industrial development has made clear its harrowing extent, though this evil might have been predictable given the cultural and technological backwardness of the countries involved. Europe and the USA contributed to this in no small part, transferring their "dirty" production to the territory of former colonies without showing appropriate regard for occupational safety of the construction of waste treatment facilities. Furthermore, developing countries often serve as dumping grounds for household and industrial waste brought in from other regions. Thus, for example, electronic garbage is delivered to Vietnam, India, Pakistan, Nigeria and Ghana in exchange for a small fee. One of the largest scrapyards for decommissioned ships makes its home in the Bangladeshi city of

Chittagong. As the vessels are scrapped, toxic lead waste products litter the shoreline, and motor oil diffuses through the coastal waters.

Let's name a few more of the most infamously polluted cities of Asia, Africa and the Americas. In Hazaribagh, Bangladesh, a leather-working center, hexavalent chromium used for tanning spills into the river without any purification. Kabwe, Zambia, is surrounded by lead contamination for miles around as a result of unregulated mining over the course of the entire twentieth century. Accra, the capital of Ghana, hosts one of the world's largest electronic garbage heaps, which is largely burned in open fires. Port Harcourt, Nigeria and the Niger River Delta, polluted with oil drilling and refinery waste. The mining town of La Oroya, Peru, called the Peruvian Chernobyl. Copper, zinc, lead and Sulphur dioxide pollute the surrounding area. Acid rain has burned away nearly all vegetation, and most residents have lead concentrations in their bloodstreams at two to three times acceptable levels.

Finally, we must say a few words about India and China, bearing in mind their special place among developing countries.

The ecological situation in China is one of the most complicated on the planet, both due to demographic overfill and the mass-movement policies of the Great Leap Forward conducted under Mao's dictatorship. During that time, millions of acres of pasture went under the plow to create more farmland, and the upper reaches of the Yangtze and Yellow Rivers were stripped of hundreds of thousands of acres of forest, leading to a massive disruption of the ecological balance, soil degradation, decertification, and wider damage areas from natural disasters, especially floods. Today, over half the population lives under poor environmental conditions, and China is home to ten of the world's twenty most polluted cities (Maksakovsky 2008, book 2). China takes first place in the world for organic pollutants in the water, which has made most of the rivers unsuitable as sources for drinking water or fish. It takes second place to the USA in carbon dioxide emissions. But while motorized transport serves as the main atmosphere polluter in Western countries, in China that role is played by hydrocarbon-based power plants and industrial furnaces that run on coal, a fuel that is also widely used by average households. From this comes the problem of smog in major cities, where you cannot go out without a facemask in inclement weather. Furthermore, the Chinese market uses soap with a high level of sulfur content, which is banned in most countries. This gives rise to regular acid rain, causing great damage to farming and forest ecosystems.

While India, after gaining independence in 1948, chose a democratic path of development, it hardly managed to improve upon the Chinese model in terms of ecological stress. In this we see the effects of both the country's colonial past and the demographic explosion of the twentieth century. The world's second largest country by population, India occupies sixth place in carbon emissions into the atmosphere and third, after China and the US, in the scale of organic pollution to surface waterways. Ninety percent of this pollution comes from industrial and domestic waste of cities, much of it dumped into rivers without any treatment whatsoever (Maksakovsky 2008). As concerns carbon emissions, 35% have their origin in industrial or power plant and about 40%—in motorized transport. Power plants most often use high-ash coal, and cars—low-quality leaded gasoline. We can add to that the large-scale use of wood as household fuel, for which large swathes of forest are constantly cut down. This has led to deforestation in India at a rate of up to 1.5 mln ha per year.

In light of all this, it should come as no surprise that many developing countries have lost a significant part of their natural ecosystems. The process of destroying tropical forests continues in the Amazon, Central America, equatorial Africa, South and Southeast Asia. In some countries, such as El Salvador, Jamaica and Haiti, they are practically all cut down already. The Philippines has left a mere 30% of its original forests. That country comes in second in the total area of forest removed each year at 10.8 thousand km², following only Brazil with its 25 thousand km² per year. In relative terms, the fastest deforesters are Bangladesh, which destroys 4.1% of its forests each year, followed by Pakistan and Thailand at 3.5% each (Maksakovsky 2008, book 1). The situation with forest ecosystems rests a bit better in Africa, though in some countries there they have been destroyed almost entirely. That includes Rwanda and Burundi, thoroughly farmed countries with quickly growing populations.

Impoverishment, unemployment and widespread vulnerability to natural disaster or military conflict have all served as causes bringing denizens of the world's poorer regions to search out better lives in the more prosperous countries of Europe and North America. Huge masses of people from developing countries are taking part in this migration process. But it was not always so.

Until the mid-twentieth century, it was Europe itself that served as the hotbed of outward migration. The Age of Discovery provided the first jolt, creating a precedent for the Old World's surplus population to flow into the unconquered expanses of Siberia, Australia and the Americas. This process reached its fullest extent only in the late nineteenth century. From 1820 to 1920, over 50 million people emigrated to the United States. At various times, this was brought about by hunger (as with the Irish in the 1840s), the tyranny of monarchical and totalitarian regimes, pogroms against Jews in Russia, genocide against Armenians in the Ottomon Empire and other assorted miseries.

But now the flow of migration has gone in the opposite direction. Just as in previous centuries, it is most often a flight from overpopulation, hunger, poverty, internecine warfare and ethnic strife. As the UN Human Development Report for 2004 said, the stream of people from poor countries has provided almost all immigration to Western Europe, Australia and North America in recent decades. Today, almost one in ten residents of these prosperous regions was born outside them. Refugees make up about 9% of immigrants, having fled political repression or war (Human development…2014). In 2013 the number of migrants was 247 million, a full 3.2% of the Earth's population (*Demoscope* 2015). And the number has only risen since.

Of course, such an enormous influx of cheap labor cannot help but tell upon the economies of the countries that accept it, especially as demand increases for workers to fill the job openings sometimes referred to as "3D" (dirty, dangerous and degrading). These jobs go first and foremost to immigrants from poor countries who, lacking union representation and social protections, are willing to accept low wages. This allows employers to save money on labor costs, providing a competitive advantage for their businesses.

Particularly advantageous to employers is the hiring of illegal immigrants who make up 10–20% of labor migrants worldwide according to estimates by the International Labour Organization (ILO). Lacking lawful residency in their country of work or any labor contract, illegal immigrants are prepared to accept low wages and difficult conditions, becoming targets for the most shameless forms of exploitation (Taran 2010, p. 70–71).

Even under these conditions, the migrants' earnings throw a meaningful lifeline to their families left at home. Considering the scale of modern migration, these various rivulets of cash, on coming together, form a mighty tributary to the economies of developing countries, second in importance only to direct foreign investment as a financial stream and doubling in amount official channels of foreign aid (Glushcenko 2005). Furthermore, emigrants working abroad acquire valuable skills, training and experience, which forms a positive influence on their return, encouraging increased effectiveness and a raised level of culture.

Granted, not all emigrants return to their countries willingly. The contrast in standard of living between Europe, Australia or North America and countries of the third world is too great. If there is even the slightest chance to gain a foothold in the new country, most migrants will use it. As demographer P. Taran notes, labor immigrants from developing countries suffered first in the recent economic crisis due to terminations, lost wages and worsening labor conditions, but nonetheless a majority preferred not to return home unless threatened with forced deportation. "Even when financial rewards were offered for voluntary departure, they preferred to stay… Because the situation at home was still worse." (Taran 2010, p. 85)

This glaring contrast between economically developed and backward countries represents a major problem of world order. While strict caste boundaries divided rich and poor as a fact of life in antiquity and the middle ages, such divisions in our own time, particularly applied to entire peoples, look like social injustice in light of the ideal of full equality under the law. Why should the people of one nationality live in happiness and plenty, even in clear excess, when others must permanently suffer hunger and want?

This economic inequality gives rise to social tension and instability worldwide, and, therefore, serves as a serious roadblock in the path to sustainable development, as most politicians understand. In order to help underdeveloped countries escape the clutches of poverty, 34 of the world's most developed states now provide official development aid under the auspices of the Organization for Economic Co-operation and Development (OECD). In 2012, this aid added up to 125.6 billion dollars, corresponding to 0.29% of the total GNP of donor countries. The US, England, Germany, France and Japan led in donations (Millennium… 2013).

Another tool of economic aid to poor countries is writing down sovereign debt. In the period from 2000 to 2010, the share of receipts from exports spent on servicing sovereign debt in developing countries fell from 11.9 to 3%. Under the IMF's Heavily Indebted Poor Countries Initiative, 35 states were totally liberated from the yoke of debt (Millennium… 2013).

But financial aid is only one of the global strategies being implemented for closing the gap between developed and underdeveloped countries. Forming such strategies would be impossible without a system of objective criteria for properly judging the social and economic condition of various countries and the *quality of life* of their populations.

This familiar term emerged into wider usage among scientists internationally in the last quarter century along with proposals to give it quantitative value. For this purpose, researchers have primarily adopted the Human Development Index (HDI), developed in 1990 by a group of economists led by Pakistani Mahbub ul Haq as part of the UN Development Programme. After all, human potential is the most important economic resource a nation can have, and one of the conditions of its functioning. So, for a postindustrial society, the worker as a harmoniously developed person represents a specific value, and the costs of education, training and healthcare are considered among their most gainful investments.

While the Multidimensional Poverty Index applies mainly to evaluating the situation in developing countries, the HDI carries a more universal meaning. It provides the opportunity to evaluate the quality of life in any country through a complex points system and to rank and compare different countries and regions. If the HDI-based rating of a country stands higher than its GDP rating, that means it is efficiently converting achievements in economic development into prosperity and living standards for its population. An inverse relationship testifies to a weak link between the economic progress of a country and the interests of most of its citizens.

The Human Development Index is calculated based on three components: life expectancy, average and expected years of schooling and real per capita income— GDP per person adjusted for buying power. These three measurements are then standardized on a scale of 0 to 1, the maximum being the highest rating that any country has achieved in each area since 1980. Thus, in 2010 the highest life expectancy was 83 years, the longest expected period of schooling was 20.6 years, and the highest yearly income per capita was $108 thousand. As the minimum, imagine a "natural zero" corresponding to the lowest figures necessary for survival: A life expectancy of 20 years, no years of schooling and $163 per person. Those are the lowest figures any country has reached in recorded history.

The HDI represents a geometric average of all three indexes within the range of 0–1, calculated through a special formula. Further, based on this synthesized indicator, the countries are ranked into three or four groups: Countries with a low HDI (lower than 0.55), countries with a medium HDI (0.55–0.7), and countries with a high or very high HDI (above 0.8) (Human Development…2010). The UN Human Development report, released yearly, contains a summary with each publication. Below we have presented selected data on HDI ratings of countries for 2013 (Table 4.2).

Norway traditionally heads up the list of countries with a very high level of human development with a life expectancy of 81.3 years and an average yearly income of $65,000 per person. Rounding out the top five are Australia, Switzerland, the Netherlands and the USA. Nearly all European states belong to the 0.8+ HDI group, including the Baltic nations, along with the most advanced countries in Asia

Table 4.2 Groups of countries with different HDI levels

Place	Country	HDI
Countries with a very high level of human development		
1	Norway	0.944
2	Australia	0.933
3	Switzerland	0.917
5	USA	0.914
6	Germany	0.911
8	Canada	0.902
14	United Kingdom	0.892
17	Japan	0.890
19	Israel	0.888
20	France	0.884
49	Argentina	0.808
Countries with a high level of human development		
53	Belarus	0.786
57	Russia	0.778
69	Turkey	0.759
70	Kazakhstan	0.757
71	Mexico	0.756
75	Iran	0.749
79	Brazil	0.744
80	Georgia	0.744
83	Ukraine	0.734
91	China	0.719
Countries with a medium level of human development		
103	Turkmenistan	0.698
108	Indonesia	0.684
110	Egypt	0.682
116	Uzbekistan	0.661
117	Philippines	0.660
118	South Africa	0.658
135	India	0.586
142	Bangladesh	0.558
Countries with a low level of human development		
146	Pakistan	0.537
152	Nigeria	0.504
156	Zimbabwe	0.492
168	Haiti	0.471
169	Afghanistan	0.468
173	Ethiopia	0.435
185	Central African Republic	0.341
186	Democratic Republic of Congo	0.338
187	Niger	0.337

Source: Human Development Report (2014)

and Latin America (Japan, Israel, South Korea, Argentina, Chile, etc.) and some of the leading petro-states (Saudi Arabia, Kuwait, the United Arab Emirates).

Along with the BRICS States of Russia, Brazil and China, the high-HDI group includes the Post-Soviet countries of Belarus, Kazakhstan, Ukraine and the three Caucasian republics. But the bulk of developing states in Asia, Latin America and Africa's periphery show themselves typical of the medium-level HDI, represented by the major examples of Indonesia, Egypt, the Philippines, South Africa, India and Bangladesh, which should give a concrete idea of what "medium" means. These are poor countries with low per-capita incomes. South Africa leads the group at $13,225 per year. The average Indonesian makes $11,612, the average Indian—$6572, and the average Bangladeshi—$3580 (The World Bank Group 2017).

Countries with a low HDI, the poorest of the poor, round out the table. Nearly all are in Sub-Saharan Africa, though the Asian countries of Pakistan, Afghanistan and Nepal also fall into this group, along with the unfortunate Caribbean nation of Haiti. Even within this group, there are outliers of extreme poverty—the Central African Republic, the Democratic Republic of Congo and Niger. In order to imagine the reigning destitution of such places, it is enough to compare them with any of the leading countries ranked. The life expectancy here is extremely low—45–50 years. These countries have the most disadvantageous social climates. Less than 30% of the population can read and write, and the per-capita GDP is lower than $1000 per year. Worse, the impoverished condition of the poorest African countries is often exacerbated by destructive and frequent armed conflicts.

Thus, the efforts of the global community to reduce the distance between rich and poor, between the flowering prosperity of some countries and the hopeless backwardness of others, have failed to make practically any impact on the poorest nations of Africa. The same cannot be said of developing nations as a whole. From 1990 to 2010, the average HDI indicator rose from 0.57 to 0.68, which means the gap between developed and developing countries closed by 1/5 over that period. If you go back to 1970, the gap closed by a quarter. As the 2010 Human Development Report puts it, "On average then, living in a developing country today is more similar—at least for these basic health and education indicators— to living in a developed country than was the case 40 or even 20 years ago." (Human Development…. 2010).

Nonetheless, despite this overall positive swing, the global polarization of wealth continues. Have a look at the following numbers. In 1960, the income gap between the richest 20% and poorest 20% of the world's population stood at roughly 30:10. By 1995, this ratio had climbed to 82:1 (Fig. 4.4). Or, taking Brazil as a prime example, in 1960 the richest tenth of that country's population received 54% of the national income, while in 1995 they were making 63% (Meadows et al. 2006).

Geographer Vladimir Maksakovsky brought forth some no less profound data to similar effect. In 2008, 360 billionaires possessed an amount of wealth equal to half of all humanity's yearly income. The 15 wealthiest individuals had more money at their disposal than all African countries south of the Sahara. This divergence of economic power is now occurring with particular intensity in the former Soviet

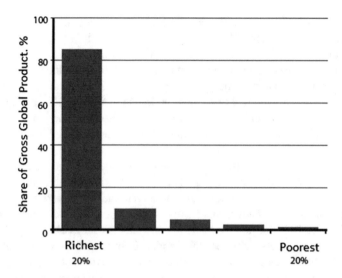

Fig. 4.4 Unequal distribution of incomes worldwide. The richest 20% of the world's population controls more than 80% of the Gross Global Product and uses 60% of energy produced worldwide. Source: Meadows et al. (2006)

republics. In Russia, for example, the number of billionaires rose from eight individuals in 2001 to 101 in 2008 (Maksakovsky 2008, book 1). That particular statistic demands no further commentary.

***

Don't think, however, that the global social crisis has bypassed economically successful states, or that a high per-capita GDP provides a guarantee against any and all social problems. Chief among such problems is the constantly smoldering tension between the native-born citizens of these countries and the "new proletariat," the migrants who have flooded many European countries since the Second World War.

On the one hand, the populations of these countries are rapidly aging, creating a sharp need for a fresh stream of labor. On the other, this process generates social tension, particularly concerning the Muslim diaspora. In France, this includes primarily Arab North Africans. In Germany, the largest Muslim immigrant group is Turks. In the UK, Pakistanis are the largest Muslim minority.

There are different approaches to fitting immigrant populations into the social structure of the host country. France uses the assimilationist model, according to which a person born on French territory, loyal to French political institutions and sharing French cultural values automatically becomes a French citizen. This is sometimes called "soil right." At its heart, this is a drive to Europeanize Islam, relegating it to the private sphere and neutralizing the distinction between the native-born citizenry and the immigrant diaspora, which loses its religious and cultural identity.

Unlike France, Germany has built its immigration model on an ethnic rather than civil principle, "blood right." According to this principle, it is not enough to be born on German territory if one plans to become a full citizen. Until recently, it was practically impossible to do so without German heritage. In 1999, the country passed a law allowing a person to become a citizen if they had been born on German territory to at least one parent who had lived there no less than 8 years. This policy of segregation and Germany's stubborn refusal to recognize itself as a nation of immigrants has led to the formation of immigrant communities isolated from the native population.

Finally, there is the pluralist model of the UK, oriented towards multiculturalism. Under such a system, the government recognizes the existence of numerous communities which have the right, within the common national society, to live in their own groupings and keep the culture and habits of their ancestral homeland. The British Nationality Act 1948 formed the basis for this, along with establishing the right of people to move freely between the Home Islands and (former) colonies as Commonwealth citizens. As a result, ethnic communities hold strong influence under the law. However, this also leads to lumping everyone with immigrant roots into separate ethnic groups with which the descendants of immigrants have little in common. Native-born British subjects have come to be viewed not as individuals of a common nationality, but as members of one ethnic community or another (Sapego 2006).

To make a long story short, let us say that despite all the differences between these three strategies, the result has turned out much the same. Whether the government sought to give immigrants full equality or made no attempt at all, there arose parallel, ethnicity-based immigrant communities. One consequence of which is Islamic radicalism, a flower of evil taken root, it seems, on alien soil.

As we know, the Kouachi brothers who committed the world-shaking terrorist attacks in Paris on January 7, 2015, were second-generation Algerian immigrants. They had been born in France. They had gone to French schools. They spoke French without an accent and made use of all the benefits of European civilization. And yet, they were not a part of this civilization. Could there have been objective, and not merely personal, causes for this break?

Let's begin with the fact that any society, no matter how advanced, is subject to social stratification. Some achieve success by birth, ability or connections, while others are doomed to occupy the lower rungs of the social ladder. First-generation immigrants, on coming to Western Europe, eagerly accepted the inglorious and low-paying jobs that corresponded to their own limited education and qualifications. Given the contrast in living standards, even these modest conditions were taken as a boon of civilization. Their children's generation, however, which took its parents adoptive country as a homeland and themselves as fully equal citizens, was psychologically unprepared to repeat their fate. But breaking the cycle of poverty is no easy task. Low social status and income, living in the worst neighborhoods as immigrants often do, constricts the opportunities to obtain good work and a quality education, sharply reducing the chances of climbing the social ladder.

Then radical Islamism comes into play, one of the political bad seeds that grow in any community. Making use of rebellious attitudes among Muslim youth who view themselves as second-class citizens, Islamists bring this social conflict into the cultural sphere. Appealing to moderate and accommodating second- and third-generation Muslim immigrants who are typically not excessively devout, they preach for walling the community off from the indigenous population and leading an insular, strictly observant life. As a result of social segregation, discrimination and high youth unemployment among French of North African descent, Turkish-Germans and Pakistani-Britons, there arises in their midst a new Islamic identity. As the Russian newspaper *Nezavisimaya Gazeta* put it, "Islam comes to dominate their consciousness, preparing them to accept any idea in Islamic guise, even the most radical" (Syukiyanen 2005).

From there, all you need is a spark. For example, in October 2005, two teenagers of Tunisian and Mauritanian descent were accidently killed when they hid from police in an electrical transformer substation, causing riots that spread from suburban Paris to Lyon, Strasburg and Toulouse over the course of two months. Rioters torched thousands of cars, looted stores, set fire to a Catholic church and caused injury to hundreds of police. The government was forced to declare a state of emergency. And this is just one of a number of similar incidents that still rock French society from time to time.

In September of the same year, 10 years before the attack on the office of *Charlie Hebdo*, caricatures of the Prophet Mohammed in the Danish newspaper *Jyllands-Posten* served as a different pretext for widespread unrest. From Denmark the protests spread to neighboring countries. However, it doesn't look like Western Europeans learned the right lessons from such events. After all, they were already hearing about how "At this time, most EU member countries do not have coherent policies dealing with this dimension of international migration" at the 2nd Stockholm Workshop on Global Mobility Regimes (Holzmann and Munz 2004). Unfortunately, such warnings were not listened to in time.

But repairing relations with the Muslim diaspora is a two-way street, and both immigrant communities and society at large would seem to benefit from a strategically sound approach. Several sectors of the European economy, including health care, agriculture and construction, are already structurally dependent on foreign labor. Meanwhile, the non-indigenous population of Europe, including Muslims, reached nearly 10% in 2013 and continues to grow due to both ongoing immigration from Africa and Asia and a high rate of natural growth. Some corners have even begun to speak of an "immigrant occupation." But for the Muslim diaspora to organically integrate into European society, that society must address it directly as is done in traditional immigrant countries like the US, Canada, Australia and New Zealand. In Australia, one in four people was born overseas. The US takes in an average of a million immigrants each year (*Demoscope* 2013). In neither case is immigration a source of permanent tension, because it is viewed as an unending process that can and should be properly focused. These countries have adopted an active immigration policy, based on service to economic interests as well as certain base principles such as family reunification, state humanitarian obligations to political refugees, etc.

In any case, the accelerating process of global migration is one of the main components of globalization, and we should accept it as a given. There is no way back, though that understanding in no sense means that its flow cannot be regulated. As a rule, the flow runs from regions at some stage of demographic explosion into countries experiencing demographic crisis or depopulation, i.e. from economically backward countries suffering an overabundance of hungry mouths and unused labor into places in need of human capital. In this way, migration eases the situation of source countries and reduces the economic, demographic and political gap between the world's regions. So, for all its shortcomings, this circulation of people ultimately serves to fortify overall stability and level the uneven playing field of global development.

As we know, lead 9/11 hijacker Mohammed Atta spent his last and somehow decisive 8 years in Hamburg, Germany's second largest city. This detail cannot be thought a coincidence. Modern megalopolises, these engines of scientific and technological progress, not only bring all the advantages and convenience of civilized life, but form or deform the psyche of those who live in them.

While a few decades ago megalopolises and heavily urbanized industrial regions like the Ruhr Valley, Donets Basin or Greater Moscow were the unique province of economically advanced countries, today the population centers of the developing world are quickly overtaking them (in population, though not in amenities). In a number of countries in the Middle East (Kuwait, Qatar, Lebanon) and Latin America (Venezuela, Uruguay, Argentina, Chile), urbanization is near or above 90% (World Urbanization Prospects 2014), which until recently was characteristic only of the most urbanized countries of Europe, Japan and some US States. However, while enabling progress in these countries and granting millions of people access to information and cultural resources, urbanization exacerbates in the extreme those social problems caused by catch-up modernization. We could then say without exaggeration that in the last 30–40 years our planet has transformed from a "big village" to a "big city," and the issue of urbanization has moved into the forefront of the ongoing global crisis.

Experts are of different minds concerning the role of major cities in the life of modern man. Some primarily see the advantages, since high population density and developed infrastructure streamlines production, concentrates the flow of information and speeds the process of innovation. Others, while not denying the drawbacks of urbanization, view it as an inevitable step in human progress. We must, therefore, in the words of Russian academic Nikita Moiseyev, "accept this reality and learn to build megalopolises in such a way we can live in them without becoming warped. And most importantly—to learn to live in these monsters" (Moiseyev 1998, pp. 50–51).

This lesson is difficult to learn, however, and for all the temptations of creature comforts, the entertainment industry or a developed healthcare system, life in a megalopolis often breaks a person on the rack in both the physical and psychological sense.

The very environment of large cities, with its high level of industrial and transport pollution and nearly non-existent facility for self-cleaning, adversely affects a person's health. If you consider that most of the territory of developed countries is located in one of the environmental destabilization centers, that these countries consume most raw materials and produce two-thirds of the world's waste, then it's not hard to imagine the role megalopolises play as epicenters of powerful disruption of the environment, which even the most advanced green technologies are powerless to stop.

In these cities, the concentration of dust particles rises to 5–15 times that of the surrounding territory, and solar exposure has fallen 10–15% over the last century. Fog or the infamous smog appears more often here, and there are 10% more cloudy days than in the countryside (Europe's Environment... 1995; Maksakovsky 2008, book 1). But most importantly, the high level of pollution in urban environments gives them many qualities harmful to the human body, which we can term their *aggressiveness*.

First of all, this concerns chemical pollution, which gives rise to 25–50% of illness in industrial centers. After all, the atmospheric emissions, runoff and solid waste from industrial cities contain thousands of tons of lead, zinc, copper, chromium and other metals. Building up in the soil and percolating into the water, here they form their own geochemical territory. Lead represents a particular danger among heavy metals. Beyond damage to the endocrine and immune systems, it also retards physical and mental development in children. A wide range of aromatic carbons possess carcinogenic and mutagenic qualities, and oxidized compounds of nitrogen and sulfur cause respiratory and bronchial illness, including bronchial asthma (Krasilov 1992). On the whole, city-dwellers suffer allergic, cardiovascular, lung and oncological disease 1.5 to 2 times more often than rural residents.

Unfortunately, aggressiveness in the urban environment does not end at chemical agents, though there has been much less discussion of other forms of physical pollution, and not a great deal of study into their effects on the human body. These include noise from transportation, which is estimated to cause $9 billion a year in damages in US cities alone, along with vibrations caused by railed transport, construction equipment, or sometimes factories. They include all kinds of electromagnetic fields ("electrosmog"), as well as ionizing radiation and any number of other physical factors significant to one's health, in which the modern citizen lives surrounded as though on a military testing ground.

Electrosmog is insidious because, like other forms of radiation, it lies beyond the human senses and its negative influence upon the body only shows itself with time. Thus, only in the relatively recent past did scientists discover the connection between electromagnetic anomalies near power lines and incidence of cancer in children. Medical researchers confirmed through study of the homes of children dead of leukemia that living in close proximity to the lines raised the chance of such illnesses two to three times (Gun 2003).

Many thousands of years ago, our distant ancestors first raised their eyes to the stars and stood spellbound to discover the immensity of the universe. Ever since, the starry heavens have never released us from their cold and silent grasp. Philosopher

Immanuel Kant said, "Two things fill the mind with ever new and increasing admiration and awe, the more often and steadily we reflect upon them: the starry heavens above me and the moral law within me." But, now it has been a full century since the residents of large cities have seen the Milky Way or the rest of the true starry heavens.

But that's the aesthetic side of the question. Artificial light pollution also strikes at a person's health, disrupting the biological rhythms of sleep and wakefulness and leaving a negative stamp upon the psyche. Night shifts at work, the glowing signboards and store fronts, night clubs and late-hour venues—all this crowds into the night and conflicts with biological human nature, devolving into mass insomnia and daytime drowsiness, along with today's "fashionable" maladies, such as depression and chronic fatigue syndrome.

While on the topic of pollutants, we must not overlook a type connected to neither chemical nor physical agents, but which has taken on ever more threatening forms in recent decades. This is *information pollution*. Modern means of communication, from personal audio players and mobile phones to television, not only raise the ambient level of sound and electro-magnetism, but are also a source of hyperinformation, the flow of which surpasses the physiological human ability to handle it by six times (Arsky et al. 1997). Lev Tolstoy, who once said that a house where songbirds are kept has no room for literary creation, could not have written *War and Peace* in our day. Worse still, in the hands of self-seeking operators, this whole information and entertainment industry is banefully deforming the minds and spirits of children, if not the consciousness of adults. Medical researchers have coined the terms computer and internet addiction. Having passed through many of the virtual battles that abound in video games, such children start to feel like supermen, and their behavior and psyche change as the distinction between real and computerized life diminishes (Gun 2003).

As concerns obnoxious television advertising, a number of psychologists and psychiatrists think that it is responsible for up to half of all growth in crime and substance addiction. "Advertising engages in psychological extortion," according to psychologist Vladimir Levi. "It puts the subconscious into junkie mode: It suggests, it implants itself, it propagates the cult of ecstasy, the ideology of getting your fix no matter what" (Levi 2002, p. 375). And here is how Vladimir Nabokov described the engrossing and hypnotic effect of advertising on a young mind in his *Lolita*:

"If a roadside sign read: Visit Our Gift Shop—we *had* to visit it, *had* to buy its Indian curios, dolls, copper jewelry, cactus candy. The words "novelties and souvenirs" simply entranced her by their trochaic lilt. If some café sign proclaimed Icecold Drinks, she was automatically stirred, although all drinks everywhere were ice-cold. She it was to whom ads were dedicated: the ideal consumer, subject and object of every foul poster" (Nabokov 1991 p. 148).

In the end, this person, spoon-fed on mass culture and psychologically dependent on invisible spirit-guides, turns into an ideal candidate, almost specially prepared for manipulation by social and political processing and the oft-referred to zombification. But this is only the backdrop, only the stage upon which big city dwellers play out their human comedy. The action in this setting, as a rule, is densely packed in the extreme, and the overwrought and nearly out of control actors squeeze in so tight as to step on each other's toes. Some authors have theorized that information about the optimal population density is stored in the human genome, and in all likelihood that code has not been overwritten (Severtsov 1992). Therefore, the constant if not always recognized discomfort of being densely surrounded which accompanies us from preschool to the grave (and often in the grave as well, as writer Aleksandr Tvardovsky remarked, "I managed to secure a tight nook in a communal apartment for eternity") cannot fail to leave its mark upon the human psyche, for all its flexibility.

Psychologists call this the crowd or group effect, when overcrowding itself becomes the cause of chronic stress and related mental problems. So everyday aggression, crime and addiction, the traditional problems of major cities, may also have a biosocial origin.

Finally, the high concentration of technological objects and means of transportation, multiplied by the extreme density of population, makes city dwellers particularly vulnerable to epidemics, accidents and natural disasters. Thus, in London in 1952, four thousand people died at once and twenty thousand more suffered injuries due to heavy smog, the worst ecological catastrophe of its kind (The World Environment 1992). Earthquakes, too, sometimes take tens of thousands of lives in cities.

The defining aspect of urbanization is the ripping away of people from their natural and cultural roots, and their frustration before the alien power of the state bureaucratic apparatus which gradually distances them from the remaining mass of citizens through administrative structures of inexorably increasing complexity.

As a result, a new structure of political power becomes the norm. Officials govern through back channels. Various breeds of image-maker and spin doctor manipulate the social consciousness. The national security state grows further and further beyond the control of the legislative and judicial powers, not to speak of society itself. As Russian philosopher and social critic Aleksandr Panarin said, "The elected 'Republic of deputies' begins to be set against the secret power of experts, and the dilettantism of public politicians—against the esoteric knowledge of professionals hiding behind the scenes, concerning the secret strings to be pulled in a dark side of politics which, on principle, shall never be disclosed" (Panarin 2000).

Under those conditions, those referred to as "plain folks" seek shelter and defense among various religious sects and other shadowy countercultural gatherings, where they try to obtain the psychological comfort they are otherwise denied. With time, however, such groups often organize themselves into authoritarian hierarchies, and, depending on the ideology and ambition of their leaders, into criminal and extremist syndicates.

Thus, to draw some conclusions, let us say that although developed countries have managed to solve many age-old social problems associated with famine, crop failure, plague and poverty that have tormented mankind for thousands of years, this was done by creating an energy-guzzling semi-artificial habitat for most of the population. Granted, this habitat provides people with a relatively safe and comfortable life, but only the most deep-seated urbanist would dare to call it healthy. Many famed psychologists from Freud on have pointed to its relative corruption and inadequacy. And it is no exaggeration to say that nearly all notable psychological prose in the twentieth century, from Kafka to Salinger and from Trifonov to Petrushevskaya, could serve as a vivid illustration to that thesis.

But there is another, more important, point. More importantly, supporting these kinds of semi-artificial conditions of existence requires developed countries to constantly expand their use of energy, which means increasing pressure on the environment that provides every last one of our kilowatt hours. Even that most tireless point of pride for the developed world, a long life expectancy, bears witness not as much to the blooming health of the nation as to added megawatts of electricity and trillion-dollar outlays to medical and pharmaceutical industries. Behind the façade of this well-being lie well-cultivated surgical and endoscopic methods, complex electronic diagnostic equipment, mountains of psychotropic and cardiovascular medication, hormones and antibiotics without which, like a junkie without heroin, the modern European, Japanese or American could never get along in their "extended" life. And, therefore, such prosperity and health, along with everything else, also comes at an ecological cost. And we must keep that in mind when we speak of the success of modern civilization.

References

Alkire, S., & Santos, M. E. (2010). *Multidimensional poverty measurement and analysis*. UNDP.

Arsky, Y. M., Danilov-Danil'yan, V. I., Zalikhanov, M. C., Kondratyev, K. Y., Kotlyakov, V. M., & Losyev, K. S. (1997). Ecological problems. In *What is going on? Who is to blame? What is to be done?* Moscow: MNEPU. [in Russian].

Danilov-Danil'yan, V. I., & Losev, K. S. (2006). *Water usage: ecological, econonomic, social and political aspects*. Moscow: Nauka. [in Russian].

Demoscope Weekly (579–580). (2013, December 16–31). Retrieved from http://demoscope.ru/weekly/2013/0579/barom03.php [in Russian].

Demoscope Weekly (641-642). (2015, May 14–17). Retrieved from http://demoscope.ru/weekly/2015/0641/barom01.php#_ftn2 [in Russian].

Demoscope Weekly (77–78). (2002, 26 August–8 September). Retrieved from http://demoscope.ru/weekly/2002/077/tema06.php [in Russian].

Europe's Environment. (1995). *Statistical compendium for the Dobris assessment*. Luxemburg: Eurostat.

Glushcenko, G. I. (2005). Cash transfers of International Labor Immigrants: Characteristics and determinants. *Voprosy statistiki*, (3), 38–50. [in Russian].

Gun, G. E. (2003). *The computer: How to keep one's health*. St. Petersburg, Moscow: Neva, OLMA-PRESS Ekslibris. 128 p. [in Russian].

Holzmann, R., & Munz, R. (2004, June 11–12). *Challenges and opportunities of international migration for the EU, its member states, neighboring countries and regions: A policy note*. 2nd Stockholm Workshop on Global Mobility Regimes. Retrieved from http://documents.worldbank.org/curated/en/158811468751785342/pdf/301600sp.pdf

Human Development Report. (2010). Retrieved from http://hdr.undp.org/sites/default/files/reports/270/hdr_2010_en_complete_reprint.pdf

Human Development Report. (2014). Retrieved from http://hdr.undp.org/sites/default/files/hdr14-report-en-1.pdf

Krasilov, V. A. (1992). *Environmental protection: Principles, problems, priorities.* Moscow: Institut okhrany prirody i zapovednogo dela. 174 p. [in Russian].

Levi, V. L. (2002). How to educate parents, or A new non-standard child. Moscow: Toroboan [in Russian].

Maksakovsky, V. P. (2008). *A geographic portrait of the world (in two books).* Moscow: DROFA [in Russian]. Book 1. Retrieved from http://www.twirpx.com/file/997779/. Book 2. Retrieved from http://www.twirpx.com/file/997899/

Meadows, D., Randers, J., & Meadows., D. (2006). *The limits of growth: The 30 year update.* London: Earthscan.

Moiseyev, N. N. (1998). Fate of Civilization. The Path Of Reason Moscow: MNEPU. [in Russian].

Nabokov, V. L. (1955) from The Annotated Lolita, Ed. Alfred Appel, Jr. (1991). *1st Vintage books* (p. 148). New York: Random House.

Panarin, A. S. (2000). *The temptation of globalism* (384 p). Moscow: Russkiy natsionalny fond. Retrieved from http://www.e-reading.link/bookreader.php/139954/Panarin_-_Iskushenie_globalizmom.html

Sapego, G. P. (2006). Immigrants in Western Europe. *Mirovaya Ekonomika i mezhdunarodnye otnosheniya, 9*, 50–58. [in Russian].

Severtsov, A. S. (1992). The dynamics of human population numbers from the standpoint of animal population ecology. Biulleten' Moskovskogo obschestva ispytatelei prirody. *Otdel biologii, 6*, 3–17. [in Russian].

Syukiyanen, (2005, November - 16). French Lesson: Europe May Become the Main Arena of a Confrontation Between the Islamic and Western Civilizations. Nezavisimaia Gazeta. [in Russian].

Taran, P. (2010). Globalization and labor immigration. *Vek globalizatsia,* (1). [in Russian].

The Millenium Depelopment Goals Report. (2013). Retrieved from http://www.un.org/millenniumgoals/pdf/report-2013/mdg-report-2013-english.pdf

The World Environment. (1972–1992). London: Chapman and Hall. 884 p.

The World Bank Group, (2011). http://web.worldbank.org/WBSITE/EXTERNAL/EXTRUSSIANHOME/NEWSRUSSIAN/0,,contentMDK:20578469~pagePK:64257043~piPK:437376~theSitePK:1081472,00.html.

The World Bank Group, (2017). http://data.worldbank.org/indicator/NY.GDP.PCAP.PP.CD.

UN Human Development Index in the World's Countries. (2011). Retrieved from http://www.undp.org/content/undp/en/home/librarypage/hdr/human_developmentreport2011.html

WWF Living Planet Report. (2010). Retrieved from http://www.wwf.ru/resources/publ/book/eng/436

Chapter 5
Centralized Economics, the Market and Their "Contributions"

In the whole period of civilization's development, humanity has truly experienced only two economic systems—market and centralized. While the former has several millennia behind it, the latter, borrowing certain elements from antiquity, underwent a full-scale experiment only in the twentieth century. With regard to environmental destruction, both of them delivered the very same result in the form of a global ecological crisis.

One might think that a centralized system, concentrating in its hands the levels of power and strictly regulating the parameters of socio-economic development, could provide all of the conditions for rational land usage and preserving the natural environment. But that most certainly did not happen. Preaching, like the capitalists, the ideology of unrestrained economic growth, Soviet leaders set the goal to "catch up and surpass" the most advanced countries in industrialization in the shortest possible time. Thus they were to develop and urbanize at the cost of merciless exhaustion of their own natural resources and disregard for the basic needs and wants of human beings. Meanwhile, it quickly became clear that in its negative aspects, the new system did not differ much from the old. It represented, you might say, an inferior, "oppositional" position in comparison to capitalism. Unlike parliamentary democracy, totalitarian socialism enabled the state to enact its grandiose plans according to the will of central authorities, without considering popular opinion or local interests. The absence of transparency and freedom of speech made it possible to hide the cost incurred by "projects of the century" and plans for the "transformation of nature," while exaggerating the true results.

Thus, the construction of the first hydroelectric stations in the Upper Volga Cascade and the simultaneous creation of the Ivankovo and Rybinsk Reservoirs led to the full or partial deluge of several cities, a great number of churches, monasteries and former aristocratic manor houses. Seven hundred towns and villages went under water, along with rich pastureland and floodplain meadows. Authorities evacuated 150,000 people from the flood zone, while 294 refused to abandon their family tombs and died (Erokhin 2003). All of this was kept secret from the citizens of "the

© Springer International Publishing AG, part of Springer Nature 2018
V. I. Danilov-Danil'yan, I. E. Reyf, *The Biosphere and Civilization: In the Throes of a Global Crisis*, https://doi.org/10.1007/978-3-319-67193-2_5

most democratic state in the world," which Lenin called, "a thousand, a million times more democratic than the most democratic bourgeois democracy."

For just as long a time, the facts remained unknown concerning two nuclear blasts used to divert the flow of rivers in Northern European Russia into the Volga basin. Furthermore, the blasts occurred ahead of schedule, without waiting for official confirmation of the project plan, as there was absolute certainty that approval would be obtained (Losev et al. 1993).

From the other side, the extensive and inefficient character of centralized economics enabled various groups and agencies to engage in wide-scale misappropriation of land and mineral resources, which formally belonged to the government, but in point of fact, to no one. The Aral Sea ecological disaster and the elimination of fragile northern nature during geological surveys and oil/gas drilling in Western Siberia bear witness to that.

In this process, we cannot fail to see historical parallels to the market economic system in the actions of the centralized type. First, in the interests of initial capital accumulation, both one and the other ruthlessly exploited and robbed the citizens of their own and other nations. Take the looting of overseas colonies, expropriation of peasant and church land in Tudor and Stuart England, the exploitation of Black slaves in America, or farm collectivization, the GULAG with its unpaid labor and the provision of bare subsistence compensation throughout the Soviet period. Then came the exploitation of nature. The Soviet economy did this on a particularly intensive and massive scale because it lacked market competition, crushed any outgrowth of civil society and held no decision-maker accountable. The party higher-ups could direct the flow of money concentrated in the State's hands toward any grandiose project it wished, enabling it, in certain preferred areas, to reach unheard of rates of industrial development. And these shock-constructed monocities, high-rise dams and huge atomic icebreakers were a consistent point of pride for Soviet leaders.

> *Just what we need, a new canal,*
> *That one could sight from Mars' surface,*

As Russian poet Alexsandr Tvardovsky once caustically noted.

But while advances in the Soviet defense industry and the creation of a nuclear deterrent somehow enabled the country to competitively struggle first with Germany and then with the United States, there was no such stimulus to other sectors and innovation declined year by year. By eliminating room for personal initiative and competition, the socialist system had unknowingly signed its own death warrant. Its execution was only a matter of time.

The time came with Gorbachev and Perestroika. As its legacy, the Soviet Empire left behind many debts. Among the heaviest debts, at least in the long term, was the ecological one.

Ecosystems within many Soviet Republics were almost completely deformed. Ukraine, Belarus, Moldova, the Baltic States and part of the south Caucasus now belong to the list of countries with few to no unviolated natural areas. The area of such preserves has sharply declined in Tajikstan, Kyrgistan and Turkmenistan.

Kazakhstan and Uzbekistan are still reaping the bitter harvest of the Aral Sea ecological disaster, which has affected all of Central Asia to one extent or another and severely compromised the natural balance of the region. Making right the consequences of the disaster, if at all possible, will require enormous expense and last more than one generation.

The Russian Soviet Federated Socialist Republic had more luck in this regard. The centralized system most intensively modernized the periphery of its empire, while for its heartland, most fortunately for it, there wasn't enough money. Even localized development required too great a capital investment under the tough conditions of the Russian north. However, ecosystems were disturbed on 35% of Russian territory (completely destroyed on 15%), an area greater in size than half of Europe. But that is the 35% where the vast majority of the country lives (Losev 2001).

There are differing views as to what caused the ruin of centralized economics. Some link the collapse of "actually existing socialism" with an overall crisis of civilization in which the market system turned out to be more flexible and, therefore, more resilient (Blanco 1995). Others suppose that the socialist system was exhausted by the arms race, to which it sacrificed more energy and resources than its Western competitors. This is undoubtedly fair. After all, lower military costs in Western countries lend themselves in part to the civilian population's mass consumption of technologies originally produced by the military-industrial complex. This expanded use provided for a rapid lowering of costs for multi-purpose production.

This was impossible in the USSR. This was due to a simple lack of resources for civilian needs, and because outlays for the military industrial complex were so high that they left no room for maneuver, and under the weight of an unbelievably strict regime of secrecy that made any transfer of technology from the military to civilian sector taboo. As a result, almost everything produced by the colossal efforts of the Soviet people—energy, ore, metal, cotton—went primarily to the defense. Civilian production received operating capital according to the "leftover principle." Therefore, Soviet defense production required a greater expenditure of resources per unit than did that of its developed, market-oriented competitors. But, the more resources the military-industrial complex consumed, the less there was left over to provide for civilian needs, and the more unit costs rose. This in turn, required ever higher outlays, and so on. In short, this was a harmful tendency with a positive feedback loop (for more on that, see Chap. 7) typical of socialist economies.

Finally, a third school asserts that the cause of the centralized system's collapse lay in a spiritual crisis of a Soviet society worn out by the chase after Communist mirages. Indeed, Soviet authorities jealously defended the paradigm of building communism. Ideologists put forward three fundamental tasks necessary to achieve this stated goal: The creation of a new system of socio-economic relations, the construction of a material and technical base for communism and the psychological formation of the so-called "builder of communism"—the man of the future.

The first of these tasks was formally completed in 1936, when the market was truly driven out of the Soviet economy and the Stalin Constitution guaranteed social rights for the population such as free health care and education, which at the time were not in the constitutions of bourgeois states. These new socio-economic relations, however, did not ultimately lead to a solution of the second fundamental problem. By the early 1960s, when authorities were forced to raise food prices for the first time since World War II, many people began to seriously doubt that it could be solved. From then on, the number of enthusiasts, those who believed in an imminent communist future, sharply declined. Vain attempts by propaganda to portray wishful thinking as reality achieved precisely the opposite of the intended result. People just stopped believing, even if they gave credit for success in education, science and some areas of cultural policy. At the same time, increasing public apathy and the impossibility of realizing one's creative nature in a totalitarian society led to its gradual degradation, deepened by attitudes of cynicism, skepticism, hypocrisy and the spread of alcoholism. By the start of the 1980s, it became unquestionably clear that the task of reforging and psychologically transforming Soviet man had gone down in flames.

Of course, as with any historical event of such a scale, the breakup of the USSR cannot be explained by a single cause. There was a whole complex of causes, including those that researchers have yet to fully study. In any case, the global defeat of centrally planned economics came to define the end of the twentieth century. The whole world looked on in awe as one of the great super-powers and its Eastern-European satellites toppled down like a maze of dominoes.

Furthermore, in the 70-year dispute, the market system claimed a most decisive victory, showing itself more flexible, more humane and, ultimately, more resilient than its counterpart. Thus it is that today, not counting North Korea, the vast majority of the political, scientific, cultural and business elites in most countries have come to a consensus that humanity has but one undisputed economic system, the market, which is fated to carry the weight as we confront the ecological challenge. But is the market up to this most difficult mission?

***

It's common knowledge that the market is blind with regard to long-term strategic aims. Its true element lies in short-to-medium term reactions to signals of current or predicted changes in demand that could affect profitability, exchange rates, returns on bonds, etc. At the same time, market indicators frequently do not reflect real changes to quality of life—public health, security, and the state of the environment—or rather, as in a hall of mirrors, represent them in distorted forms.

Thus, the system worked out under the UN a half century ago for accessing national accounts, or the even earlier method of calculating Gross Domestic Product (GDP), records an increase in measurable economic activity due to environmental pollution, exhaustion of resources, or even accidents. Indeed, whenever a car crashes, the cost of medical care for the victim, or his burial, insurance payments to his relatives and the purchase of a new car are all factors of measurable economic activity as goods and services which, however paradoxically, contribute to GDP expansion. On the other hand, breast feeding a child does nothing to enable GDP growth, but feeding her formula out of a bottle makes a certain contribution to economic vitality. Or, let's say you get a coronary artery bypass graft—that is simply manna from heaven, falling on medical and insurance companies.

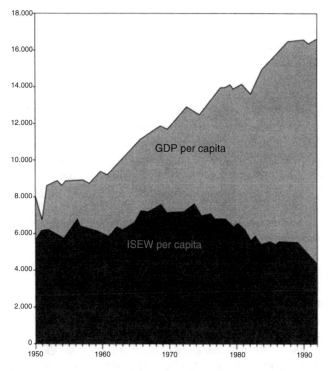

Fig. 5.1 Comparison of trends in GDP, above, and Index of Sustainable Economic Development, below, per capita in the USA from 1950 to 1990. Vertical axis—trillions of dollars according to 1952 price index. Source: http://www.tronland.net/cryptron/crypt-e.htm

In order to make up for this obvious contradiction, senior economist of the World Bank's environment department, Herman Daly, and John Cobb, Jr. proposed to bring a new account system into usage alongside the GDP—the Index of Sustainable Economic Welfare (ISEW). Measured in the same monetary units as GDP, this indicator allows us to count typically unnoticed ecological and social factors, such as domestic labor or devaluation of natural capital, and reflect a different socioeconomic cross section including quality of life factors (Weizsacker 1998). Correspondingly, the social organism looks very different in this mirror. From the late 1960s, for example, GDP rose significantly, while the Index of Sustainable Economic Welfare demonstrated a clear downward tendency in a number of countries (Fig. 5.1). This divergence once again confirms the reality of conflict between business interests and the interests of the great bulk of the population.

The image of the marketplace as the embodiment of democracy and freedom has imprinted itself upon the public consciousness. But meanwhile, even in the most progressive democratic states, firms and corporations generally function according to a strict authoritarian model. Employees do not choose the leadership and take no part in making vitally important strategic decisions. As philosopher and social critic Aleksandr Panarin noted, "The monopoly of private property and the laissez-faire position of social channels, whether external (outside the business) or internal (inside the business), are considered a guarantee of economically rational behavior, oriented towards the maximum possible profit" (Panarin 2000). As a means of cushioning this authoritarian model, a number of major companies in the US, EU and Japan have brought in systems of participatory management. These systems propose involving personnel in discussions of problems facing the company and possible methods of solving them. However, participatory management in practice allows top management to more fully use the abilities of personnel without relinquishing to them any additional powers or obligation to follow the recommendations.

It works out that a person labors through their shift under an authoritarian system and returns to democratic society only upon leaving the company gates. The West could not resolve that contradiction. Even beyond the company grounds, however, the renowned freedom of choice developed countries take such pride in often limits itself to consumer preference for one of those very same corporations. The market transforms itself into a mechanism to create and shape demand, including demand that goes beyond reasonable human wants.

Market products, wrote philosopher and social critic Herbert Marcuse, whether mass media, household goods, clothes, food, the inexhaustible array of entertainments or the information industry, "indoctrinate and manipulate; they promote a false consciousness which is immune against its falsehood. And as these beneficial products become available to more individuals in more social classes, the indoctrination they carry ceases to be publicity; it becomes a way of life" (Marcuse 1964). And "A study for the US National Academy of Engineering found that about 93 per cent of the materials we buy and 'consume' never end up in saleable products at all. Moreover, 80 per cent of products are discarded after a single use, and many of the rest are not as durable as they should be" (Von Weizsacker et al. 1998).

Many firms don't even make a secret of their vested interest in having products serve the consumer for only a limited period and break down soon after the warrantee expires. As a rule, repair is impractical and expensive. American and European garbage dumps full of discarded but entirely functional home appliances bear profound witness to this. And all of this is an entirely legal way to activate consumer demand, enabling growth of GDP.

Once, at the dawn of liberal economics, Adam Smith looked upon a sense of self-command among individual producers as an irreplaceable regulator of civilized market relations. Much water has since passed under the bridge, and, as the number of producers has multiplied and multiplied, the previous moral restraints based on religious (especially Protestant) values have been materially transformed. The market today is different in many ways from that of Smith's day. Shameful trades from the margins of business feel ever more sure of themselves. Gambling, the sex industry and drug peddling prosper in the exploitation of human weakness and vice, turning them into a colossal source of revenue. A completely new, before unseen understanding of hyperconsumption has emerged. In the withering drive to corner the market, the attempt by one party to surpass rivals and drive them out of the game leads to a buyer's panic among the others, and with it a well-known psychological dependency, the feeling of being drawn upon the rack of big business.

Thus we have, for example, the compulsive desire to make purchases even to the detriment of the family budget, known to psychologists as oniomania and popularly as shopaholism, when a person in a one-stop shop forgets all notions of their own needs and abilities and proceeds to buy one thing after another. Oriented toward this end are the regularly organized discount sales, which count on initiating such behavior. According to the Frankfurt magazine *Neue Zeiten* (2007, #79), about 15 million shopaholics live in the US alone. Among Europeans, between 2 and 10% of the adult population suffers a pathological drive to shop. And this is but one avenue for the market's thoughtless squandering of resources, many of which, such as natural wealth, cannot be measured in dollars and cents. Spread out among everyone, we hardly notice this loss at all.

***

Those shortcomings in modern consumer society demonstratively illustrated themselves in the course of the most recent financial-economic crisis, which began, as you know, in the US in 2008. So, what happened that fall of 2008?

Let's begin with the fact that, until relatively recently, the capitalist world knew mainly crises of overproduction. The best known of these, the Great Depression, also broke out first in the U.S. In recent decades, the structure of real economies of developed countries has changed deeply. High-tech production has taken a leading position, squeezing out the industrial, agricultural and raw materials sectors. There is a deeper, hidden meaning in the distinction we apply to today's post-industrial or information society, which is based on this new economy.

Indeed, whenever the web of technology grows more complicated upon the introduction of a new product, it increases the relative weight of the information component. No product escapes this process, even something as straightforward as a hammer. Its grip, made to comfortably fit the human hand, or its head, absorbing the shock of the blow—all of this contains information about how to hold and use the tool. The higher a society rises on the scientific ladder, the more information-saturated the products it uses. Just compare a thatch hut to a modern apartment block built according to architectural design and encapsulating the whole spectrum of engineering breakthroughs. You can divide the costs of any intermediate or final product between those of extended natural materials and those of added information. As a society moves up the technological ladder, the relative cost of the latter component rises, while the share of natural, raw materials costs declines.

An economy produces not only material benefits, but intellectual products as well. While these also expend certain raw materials (at least energy), the costs are so much lower structurally that one might consider these products purely informational.

In Table 5.1, we have presented the GDP structure of twenty countries for the years 2011–2013. As you can well see from this, the service sector grows depending on the type of economy and level of economic development. It is largest in advanced countries such as the U.S., Germany and Britain, followed by the approaching Poland, Brazil, Turkey, etc. On the other hand, you have economically underdeveloped states like Egypt, India or Kazakhstan, where primary (resource-exploiting) and secondary (processing) economic sectors stand equal or higher than the service

Table 5.1 GDP Structure of 20 countries, 2011–2013

Country	I, %	II, %	III, %
Australia	3.6	21.1	75.3
Angola	8	67	26
Belarus	9.5	46.1	44.4
Brazil	5.5	28.7	65.8
China	17	49	34
Egypt	13.5	30.5	56
Estonia	3.9	29.7	66.4
Germany	0.8	28.1	71.1
India	14.4	27.9	57.7
Japan	2	36	62
Kazakhstan	5.2	37.9	51.9
Luxembourg	0.4	13.6	86
Poland	4.5	31.2	64.3
Portugal	2.6	22.2	75.2
Romania	7.5	33	59.5
Russia	4	36.2	59.8
Singapore	0	26.6	73.4
Turkey	9.2	24.7	66.5
United Kingdom	1	23	76
United States	0.9	19.7	79.4

I—Agriculture, fishing, logging
II—Industry, including milling and construction
III—Services
Data from internet sources

industry proportionally. But in outlier countries, such as Angola, the relative share of the service sector gives way to the first two by several times. These distinctions bespeak a difference not only in geography but in history, as each country rests at a particular stage of socio-economic development. Thus Angola, figuratively speaking, is somewhere near the level of Europe in the late eighteenth and early nineteenth centuries, while the economies of Egypt or India correspond to levels of development in European countries in the first half of the twentieth century.

Of course, it is mainly progress in STEM fields (science, technology, engineering, math) that holds responsibility for the information saturation of society. Once the genie is out of the bottle, however, there's no predicting what it might work. There's no questioning the benefit where the service sector has served the interests of comprehensive human development—easing people's everyday lives, improving their health, providing cultural enrichment and educational advances. Unfortunately, the most important segments of the information economy—the leisure industry and financial sphere (stock transactions, bond trading, credit services, etc.)—have developed lives of their own, narrowly if at all bound by humanistic values, growing into autonomous self-supported and self-expanding structures. (For more on this, see Danilov-Danil'yan 2015.)

These industries drew in surplus financial resources, given that they made money quickly and easily. Often they required almost no effort at all, though they came with a high degree of risk. For some, the risk was simply taking a loss on the transaction. For others, it was losing a revenue stream due to the fickle nature of tastes and fashions. Never before, however, did surplus consumption ever cause a financial crisis on such a scale all by itself. For that to happen, it needed to become a mass phenomenon. That's just what happened in the final third of the previous century, as major strata of the population in developed countries obtained impressive sums of money, and with it the resource of free time, being unprepared for this in culture or ethics. Older generations should remember the 1960s–70s slogan, "The American way of life," which the American propaganda machine used even more often than the words "freedom" and "democracy." This "way of life" assumed the prosperity of the "common man" within a consumer society based on market economics. But while it provided a high standard of living, the developed market economy did practically nothing for the cultural enrichment of the average American or European.[1]

Just the opposite, brainwashing by means of advertising and other PR techniques turned out to be more profitable, primitivizing human spiritual longings and debasing high culture to low with cinematic blockbusters, titillating romance and mystery series, pop music, fitness clubs, professional sports, casinos and so forth. A hundred thousand seat stadium brings in incomparably higher revenues than a philharmonic, where the concerts require philanthropic support. And if you set up a stage on the football field and make those same hundred thousand sing and clap along to some trite refrain, you can increase the revenues still further. Marcuse called these sorts of directed wants among the "common man" oppressive, since the social environment around the individual is formed to allow them personally no other choice. There is only one way-to live "as it should be" in a consumer society. "Individuals identify themselves with the existence which is imposed upon them and have in it their own development and satisfaction" (Marcuse 1964).

Under these circumstances, when this low mass culture floods the world, eroding national traditions and hallowed moral values, the financial and entertainment segments of the service sector transform into demons of the marketplace, destabilizing worldwide economic space. When hard times come, the people, sated on "bread and circuses," for understandable reasons, will first turn away from circuses. And since the service industry and consumer goods production has grown into such a large role in the modern economy, employing many millions of people, and their incomes provide demand for goods and services in other areas, clearly a drop in demand will result in an economic crash and subsequent insolvency for a large number of debtors and the economic agents linked to them. Therefore, an economy based on consumer society cannot be sustainable (for more on this, see Danilov-Danil'yan 2009).

***

---

[1] Many of the formulations concerning this issue were developed over a century ago by American economist and sociologist Thorstein Veblen in his book, *The Theory of the Leisure Class* (1899). See (Veblen 2009).

Nonetheless, it would be too great an oversimplification to assert that the market system always operates at cross-purposes to life, as many communists, among others, try to present it. Or that it conflicts with genuine human needs. Ultimately, in the natural environment as well, the preservation and sustainability of species is provisioned by nothing other than competition among organisms, which serves as the foundation for the multitudinous structure of life on earth. By no coincidence did noted Russian ecologist Viktor Gorshkov choose "The Biosphere as 'Free Market'" as the name of a chapter in his monograph *Physical and Biological Bases of Life Stability*. And even Adam Smith's "invisible hand," which harmonizes the needs of society with those of countless market players, each aspiring to the achievement of their own aims, is essentially an embodiment of general life principles and how they are linked to the requirements of civilization.[2]

At the heart of the market lies the activity of real people following their personal interests, and equally the interests of the social groups they represent. By entering into business relations with each other, they inevitably come to a principle of competitive interaction, perhaps even encoded at the genetic level. Meanwhile, no mathematical model or computer program is capable of providing, for example, the highly precise parity of value which, under no external influence, the market itself supports. It was ignoring exactly this natural mechanism organically ingrained in humanity on the basis of distinct private interest that ultimately led to the collapse of centralized economies.

Indeed, private initiative and the competition of free producers have already demonstrated their enormous potential, and there is hardly a need to prove that, without this powerful engine, the progress of the modern age would simply have not occurred. But we cannot fail to take into account alongside this the selfish and mercenary motives that lie at the heart of any private enterprise, which lead to unpredictable and sometimes ruinous consequences and which, given the looming ecological threat, arouse particular worry. Clearly, this situation demands a review of certain important ground rules.

The previously mentioned economist Herman Daily conceived of a few simple rules concerning use of the natural environment back in 1990, which would allow the market economy to function effectively without undermining sustainability. These rules appear more or less obvious today, and yet not one of them is enforced at the global level or even in a single country if one takes imports into account. Here are the rules:

1. For renewable resources. The rate of use should not exceed the rate of restoration for any resource. Thus, the cutting down of forests should not exceed their natural growth and the rate of forest restoration. The catch of fish—the ability of

---

[2] Granted, it is not clear why competition in a biota does not lead to the destruction of long-term stability, as we often observe in human societies. The solution to important global problems rests on answering that question, including how to provide sustainability for a civilization based on the principles of competitive (market) interactions. If anyone can manage to work out that puzzle, it would put us much further on the path to sustainable development.

schools to reproduce themselves. Water usage should not violate the mechanisms through which water resources are replenished.

2. For nonrenewable resources (fossil fuels, metal ore, etc.). The rate of use for any nonrenewable resource should not exceed the rate at which it is replaced by another, renewable resource. For example, the sustainability of the oil industry is possible only on the condition that part of the received revenues be directed toward the improvement of wind and solar generators, or toward the development of technologies that use other renewable energy sources, so that at the moment of oil reserve exhaustion, a suitable replacement will be prepared.

3. For pollution. The rate of its creation should not exceed the rate at which the polluting substance can be dissipated, absorbed or reprocessed by the environment without causing harm to the corresponding ecosystems. For example, wastewater can be dumped into a river, lake or pumped into the groundwater horizon only on the condition that the bacteria and other organisms that participate in processing them are able to keep up with the stream of substances that feed them without violating the balance of water ecosystems (Daly 1990).

The question, of course, is how well these limits fit into established economic practice. The market system as it exists today, after all, is only capable of perpetuating the tangle of contradictions to which it is irretrievably bound—wealth and poverty, hypertrophic consumption, a prodigal regard for resources, pollution of all kinds and unceasing encroachment upon nature.

Meanwhile, the ecological costs of civilization are becoming economic costs in ever more apparent ways. They have already brought about a downward trend in global economic productivity, and the now accustomed economic growth is restrained not by the availability of financial, labor or productive resources, but by the state of the environment. Whereas, in the past, one of the main limits to growth was access to man-made capital, Daly notes that today, after an unprecedented increase of that capital, it has been replaced as a limiter by "natural capital." Not the number of fishing boats, but the reproductive capabilities of fish. Not the number of farm laborers, tractors and combines or chemical stocks, but the amount of fresh water for irrigation. Not the facility of a supply chain to cut and process timber, but the quantity of forest that remains, etc. And the global economy is just now coming face to face with human-caused global warming and other climate changes as chief among its limiters, though a full assessment of these consequences still lies ahead.

***

In one of the summary documents from the 1992 Rio de Janeiro Earth Summit (more on that in Chap. 8), it says that the global path of development as a means of satisfying the growing needs of humanity have begun to deeply conflict with the environment. In order to move away from danger, it is necessary to direct this development down a safer and more orderly road.

Up to this point, economic development has taken an undirected spontaneous character, which made the capitalist system a target for criticism from Marx and his followers. These juxtaposed it against a rationally constructed classless society of the future, which they imagined would function according to a single plan coordinated from a single nerve center. It took 70 years to finally lay bare the illusions of this experimental project. Contrast this with the recovery periods of Weimar Germany and New Economic Policy Russia in the 1920s after the destruction of war and revolution, or the "Economic Miracles" of Japan and Germany in the '50's and '60's, when those countries rose like phoenixes from the ashes after catastrophic defeats in World War II. The impressive vitality demonstrated by market economics, its ability for self-healing even in the most critical situations, has finally secured its right to a monopolistic position in the global economy.

But, can we go on looking to the market as a panacea for any and all evils that might be lying in wait for the human race? Is it able, in this case, to lead humanity away from danger? There remain more questions than answers, and most of them revolve around the global ecological crisis, the most important crisis of modern civilization. It's hard to deny that in its time the market economy unleashed the process of environmental destruction on the planet. Granted, it was the flagship countries of the market system that first recognized the full seriousness of the ecological threat and put sustainable development on the agenda.

There comes the rub. The more palpable the issue of realizing this unprecedented project gets, the more it comes under practical scrutiny. For one thing, it's becoming more and more obvious that these difficulties are built into the specifics of the market economy itself, among its inherent features. In order to rise above corporate and narrow state interests, above market selfishness and concerns for short-term gain, the market must get over itself and get over several basic elements of its 400-year existence. That is what the transition to sustainable development requires.

With that in mind, let us turn our attention to the ever more frequent pronouncements (and not only from pro-communist circles) questioning the very basis of the market economic system—the paradigm of free, undirected development. Thus the authors of the Club of Rome's second report, *Mankind at the Turning Point*, Mihajlo Mesarovic and Eduard Pestel, came to the conclusion through use of computer models over 40 years ago that spontaneous global economic development under current circumstances was not only irrational but dangerous (Mesarovic and Pestel 1974). It does give you something to think about.

Private enterprise, after all, is based on the competition of free producers and, therefore, acts spontaneously by its very nature. And like any largely spontaneous natural phenomenon, it carries a force at once both creative and destructive. One could look at the regular overproduction crises that struck the Western world right up to the 1930s. The cure to this ailment came only with the induction of stricter rules to business.

Today, however, we witness not an internal, but an external crisis of "overproduction" by civilization itself. Like the "Magic Porridge Pot" in the Brothers Grimm fairy tale, it is now producing too much for our planet, pouring over the sides, streaming through houses and streets and covering the whole environment. Can modern civilization, while keeping its "spontaneous" quality, learn moderation and restrict itself? It's worth remembering that in the fairy tale it was an outside actor, the girl who owned the pot, that managed to run home and yell the magic words: "One, two, three—cook no more for me!"

To carry the analogy further, it would be illogical to escape the fact that the market economy, under the conditions of ecological threat, also needs somebody standing outside the system who could establish external limiting parameters. The market requires that these external restraints, while not suppressing its active, life-giving force, allow future development along more rational, organized and safer lines. As Daly noted on this account, the mechanism for distribution of resources and sharing of revenue by the market can be assigned prescribed ecological and ethical limits "The market is not free to set its own boundaries, but it is free within those boundaries" (Daly 1977).

Here we ought to remember that there have already been similar precedents in history and that liberal economies have functioned perfectly well under the conditions of strict state, that is to say, externally established, regulation. Granted, many examples, aside from the New Economic Policy, came as a byproduct of the establishment of a state of war. Thus, the United Kingdom created a centrally regulated market economy during World War II. This not only dealt wonderfully with the demands of defense under German blockade, but managed to provide the home front with an entirely adequate standard of living, in sharp contrast to even the most secure regions of the Soviet Union with its planned and collectivized enterprise. In Japan, during the same years, owners of major businesses were practically stripped from power. No nationalization took place—management was simply handed over to bureaucrats.

Today, however, in the face of global degradation of the environment, when the fate of all civilization worldwide stands in question, discussion of centralization needs to go to another level. Because the biosphere, a unified and indivisible resource for humanity—more accurately, for all biological species living on Earth— cannot be disarticulated into regional and state compartments, though we observe such pretentions constantly. Most countries, by tradition, view their natural surroundings not as an inseparable element of the biosphere, but as unchallenged sovereign property. Such conceptions made sense half a century ago. One might guess at whose interests are truly served by the enactment of an "independent" ecological policy. All of this, of course, has a weak connection to the understanding of sustainable development.

If we are to face the problem head on, with all due responsibility, we must acknowledge that an international, supranational organ vested with the corresponding powers that could enact a program of global environmental stabilization is as necessary to the world community as oxygen itself. It would enact this program first of all in the interests of the biosphere (and, hence, humanity as a whole), and not of any separate geopolitical, ethnic, confessional, corporate or other entity. Call it a World Government, or what you will, but let us briefly sketch out a few of its functions.

For example, this organ would conduct ecological monitoring and exercise regulatory control over the state of the Earth's environment, a bit like a planetary version of the Russian Hydrometeorology and Environmental Monitoring Service, and endowed with significantly greater powers than the current UNEP. It would hold veto power over technologies and projects contrary to the interests of global stability, like the Russian *Ekoekspertiza* or the American *Environmental Protection Agency*. It would develop social, ecological and economic scales for sustainable development that carry the power of law, a planetary *Komstandart*. It would plan recultivation of land and restoration of natural ecosystems by country and continent. It would establish quotas for energy usage, greenhouse gas emissions and other global-scale pollutants. This organ would determine the amount of aid to be allocated from a global fund to underdeveloped countries in the throes of food, demographic or ecological crises.

And let's be realistic: today's global community is clearly not ready for such thoroughgoing measures. They presuppose a voluntary waiving of a real portion of national sovereignty. Sovereignty is held sacred and makes up perhaps the sorest point in international relations. Though, as Club of Rome founder Aurelio Peccei wrote on this score, "the principle of sovereignty comes in handy to the ruling classes, which are its most strenuous advocates. The sovereign state is their fief. Its pomp, pageantry, rhetoric and glorified egocentrism coupled with the vested interests inherent in it suit them perfectly" (Peccei 1977, p. 164).

English historian Arnold Toynbee put it more categorically: "the intensity of worship of the idol of the national state is, of course, no evidence that national sovereignty provides a satisfactory basis for the political organization of mankind in the atomic age. The truth is the very opposite … in this age, national sovereignty spells mass suicide" (Quoted in Peccei 1977). And make note, this was written long before the scale of the current global crisis was recognized.

The democratization of society provides a certain inoculation against these types of "National Obsessions," without which we would not have witnessed the birth of a united Europe. There are, however, stumbling blocks here as well, which might come into play during the transition to sustainable development. Sustainable development, after all, inevitably demands a certain degree of self-limitation from the populations of developed countries. Otherwise, it would be impossible to reach any acceptable compromise with nature. But to retreat from the standard of living one has attained, even if it means doing away with many obvious excesses, is no simple task psychologically. Particularly difficult is the fact that these steps are not dictated by such a convincing motive as war, economic crisis, natural disaster or any other, more immediate, threat.

With this in mind, the ideals and values of sustainable development are unlikely to quickly attract an overwhelming majority of the population. More likely, these principles will initially remain the province of what you might call an enlightened minority. This means that any national government supporting potentially unpopular measures—limiting energy usage, for example—inevitably is held hostage by the "unenlightened majority," who will throw them out in the next presidential or legislative election. As Jacques Attali, former head of the European Bank for Reconstruction and Development, justly noted, in a democracy where leaders cannot allow themselves temporary unpopularity, bigger problems may arise in the future: "How can one think of perspective if they are always checking the polls? An inability to think of the future and take risks for its sake is a rejection of development" (Quoted from Sabov 2007).

In this sense, that same democratic government would unquestionably stand to gain from delegating the most vulnerable part of its powers in this respect to an impartial supranational organ, which could sort out the thornier issues of the ecological crisis without kowtowing to interest groups or corporate lobbies. These latter, of which we should be well aware, will probably always conflict with strategic global goals and the establishment of sustainable development to a certain extent. Therefore, the global community could hardly get by without this new authority possessing a strong hand. As Russian geographer B. B. Rodoman noted on this score, democracy in the social and economic spheres could get along perfectly fine even with totalitarianism in the resolution of ecological issues (Rodoman 2004). Though, to be clear, totalitarianism is not what anyone is talking about here, but merely the necessary uncompromising strictness in questions of maintaining environmental stability, which is directly tied to humanity's survival. And, naturally, if every country follows this path voluntarily, it is the same as adopting a mandatory strategy.

Note as well that the transition to sustainable development does not in any way presuppose the elimination of the market system as such (which most totalitarian regimes have blundered into), but only the reorientation of the world economy and its subordination to the needs of global stability. And while it is impossible to reach this goal purely by economic measures, we cannot underestimate the role they have to play.

The works of English economist Arthur Pigou (1877–1959) arouse particular interest in this respect. He was one of the first to call attention to the divergence of private and social interest (social benefit) in instances when the market does not take the results of an economic activity into account (as with environmental pollution) or assesses them inaccurately, either higher or lower. One might consider the example of the market's overvaluation of primitive mass culture, which exerts a destructive influence on the modern person's spiritual and intellectual well-being. Pigou referred to these side effects of economic activity as externalities, and proposed a special, corrective tax on businesses that produce specific social costs in order to neutralize them. In cases where the social benefit exceeds the private, such as environmentally-sound land usage, the government should support this activity with the aid of corrective subsidies.

Remarkably, Pigou came to the conclusion about a century ago that the "free market" system gives rise to conflict not only between private and social interests, but within the latter also between momentary benefit and the interests of future generations. Therefore, the government should use its mechanism of income redistribution not only to do all that it can to improve social well-being, but also to support the sciences, education, health care and the like in defense of the interests of the future (Pigou 1920).

Nonetheless, humanity will clearly not get along without the coordinated efforts of the global community, directed in part toward the creation of stricter ground rules for business. Let's not mince words: this is a very complicated and painful issue. But the life and fate of the next generations depend on providing effective regulation over the development processes of our common civilization for the benefit of ourselves and our natural surroundings.

References

Blanco, J. A. (1995). The Third Millennium. *Latinskaya Amerika, 9,* 4–14. [in Russian].

Daly, H. E. (1977). Steady-state economy: The economy of biophysical equilibrium and moral growth. San Francisco.

Daly, H. (1990). Toward some operational principles of sustainable development. *Ecological Economics, 2,* 1–6.

Danilov-Danil'yan, V. I. (2009). A global crisis as a result of structural shifts in the economy. *Voprosy economiki, 7,* 31–41. [in Russian].

Erokhin, V. I. (2003). *Russian Atlantis. A guidebook through the flooded cities of the Upper Volga.* Pybinsk: Format-print. 48 p. [in Russian].

Losev, K. S. (2001). Ecological problems and prospects for sustainable development in Russia in the 21st century. Moscow: Kosmosinform. 400 p. [in Russian].

Losev, K. S., Gorshkov, V. G., Kondrant'yev, K. Y., Kotlyakov, V. M., Zalikhanov, M. C., Danilov-Danil'yan, V. I., Golubyev, G. N., Gavrilov, I. T., Revyakin, V. S., & Grachkov, V. F. (1993). In V. I. Danilov-Danil'yan & V. M. Kotlyakov (Eds.), *The ecological problems of Russia* (p. 390). Moscow: VINITI. [in Russian].

Marcuse, H. (1964). *One-dimensional man.* Boston: Beacon. Online version: http://www.marcuse. org/herbert/pubs/64onedim/odmcontents.html Part I, Chapter 1.

Mesarovic, M., & Pestel, E. (1974). *Mankind at the turning point* (210 p). New York: Dutton.

Panarin, A. S. (2000). *The temptation of globalism.* Moscow: Russkii Natsional'ny fond. 384 p. Retrieved from http://www.e-reading.link/bookreader.php/139954/Panarin_-_Iskushenie_globalizmom.html [in Russian].

Peccei, A. (1977). *The human quality.* Oxford: Pergamon.

Pigou, A. (1920). *The economics of welfare.* London: Macmillan.

Rodoman, B. B. (2004, November 4). Russia—an administrative-territorial monster. POLIT. RU. Retrieved from http://polit.ru/article/2004/11/04/rodoman/ [in Russian].

Sabov, D. (2007, Jan 21). The New Frenchman. Ogonyok. Retrieved from http://www.ogoniok. com/4979/20/ [in Russian].

Veblen Torsten, (2009). The Theory of the Leisure Class. N.Y.: Oxford Univ. Press. Online: http:// moglen.law.columbia.edu/LCS/theoryleisureclass.pdf

Von Weizsacker, E., Lovins, A. B., & Lovins, L. H. (1998). *Factor Four: Doubling wealth, halving resource use.* London: Earthscan.

Chapter 6
Humanity's Spiritual Crisis as the Root Cause of the Ecological Challenge

Thus, by the start of the third millennium, global civilization had entered a state of deep systemic crisis whose social, demographic and economic aspects, though serious in their own right, pale in comparison before the main, ecological threat. At the center of this process of global environmental degradation that overtook practically the whole planet in the twentieth century sits the culprit—man and his consumer relationship toward nature, his warped psychology of conqueror and lord of the universe.

This psychology, as we have noted, did not form yesterday. The first milestone on this path of detachment from nature came when humans transitioned from simply accepting its gifts as primitive hunter-gatherers to agricultural production, the work of our own hands (the Neolithic Revolution). But, for that reason, those very hands now hold the fate of their owners.

Against this backdrop of relative independence from natural forces, humans have lost their previous sense of kinship with the surrounding plant and animal world. Thinking himself a special being, chosen among all others, man delayed little in discarding his zoomorphic gods, who had laid down more than a few commandments against the extermination of living nature. In their place came gods that man had made in his own image and likeness. Likewise, nature itself turned from an object of worship for man into his resource storehouse.

This rationalized consumer relationship toward the environment was the norm even in the early civilizations of antiquity. The first civilizations, such as those in Mesopotamia and the Mediterranean, all evolved according to a single, repeating scenario, here noted by paleo-biologist Valentin Krasilov: "population growth—exhaustion of resources—expansion—militarization, totalitarianism—overburdening of resources (Italian forests went to constructing the Roman Fleet)—degradation of living space—spiritual degradation—loss of internal energy—collapse of the system of government" (Krasilov 1992: p. 14).

In this way, civilized humanity formed the conviction in early antiquity which Francis Bacon later phrased, "Knowledge Itself is Power," and that it could be used to adapt and remake nature for one's own needs, simultaneously revolutionizing the

© Springer International Publishing AG, part of Springer Nature 2018
V. I. Danilov-Danil'yan, I. E. Reyf, *The Biosphere and Civilization: In the Throes of a Global Crisis*, https://doi.org/10.1007/978-3-319-67193-2_6

social and economic conditions of life. Of course, it would be difficult to object to Bacon's maxim were man in his self-confidence not wont to mistake his incomplete knowledge for absolute truth.

After the "Dark Age" of the early medieval period, when Nature assumed a more protected role as "God's creation" (Medieval European principalities had numerous regulatory constraints), rationalism was reborn under new conditions. The powerful impulse for modernization in European society gave rise to the Renaissance, the earliest stage of our own Modernity.

This period is marked by the formation of modern nation-states, as well as the institution of market systems and liberal civil society in much of Europe. In the spiritual sphere, Modernity brought about what amounts to the second revolution of worldviews, after the Neolithic one. Along with sailors, writers, architects and artists, an enormous contribution to this process was made by scientists—Descarte, Gallileo, Newton, Adam Smith and others. Under this worldview, man stood in the center, at the pinnacle of creation, with increasing belief in his might and the power of reason, which he could use to transform and improve the Earth.

> *Two worlds belong to mortal man:*
>   *The first did us creation give,*
>   *The second one, in which we live,*
>   *We've built through ages best we can* (from "Na zakate", 1958),

the Russian poet Nikolay Zabolotsky said. And truly, the great geographic discoveries, the modernization of sailing fleets, the growth of cities, the establishment of manufacturing, etc., it would seem, confirmed the uncontested superiority of that man-created world. At the same time, unconquered, primordial nature fades into the background of our consciousness, understood only as the setting for our grandiose exploits.

By no coincidence, it is most often as a backdrop that nature appears on the canvas of Renaissance masters. Though not without its charm, it appears conventional and detached. In contrast to the poetry of antiquity, literary works of the time feature practically no description of nature.

The baroque and neoclassical eras and the flowering of garden design and landscape architecture, perhaps, cracked open a window for nature into the human world, but only for that which people of the time wanted to see—"regular," groomed, cultivated—and again as a setting for palace soirees and amusements. The stamp of "adaptability" in depicted nature to human demands is printed upon the work of famous painters in the seventeenth and eighteenth centuries. A sympathetic, often awe-struck view on the world of "useless" wild nature arrived even later, at the end of the nineteenth century, after nearly a century-long reign of the romantics in European arts.

While in many ways still conventional and romanticized, this nature suddenly seemed concordant with the human heart. It seemed important for its own sake, as it was independent of any practical advantage that might be taken from it. The band began to widen, starting with artists (Constable and Turner in England, the Barbizon School and impressionists in France, Levitan and Kuindzhi in Russia) and continuing among writers (Melville, Thoreau, Thompson Seton, Prishvin, later Pasternak). They drew inspiration from nature, and there sought answers to many troubling questions of everyday life.

But that was art. In reality, in everyday life, everything was going differently. The ultimately victorious free market with its all-encompassing relations of monetary exchange dictated its own logic, and that logic was the Almighty Dollar. Wherever there was a choice between leaving nature untouched or extracting an additional profit from it, man executed the latter option without hesitation, never thinking of tomorrow. We reduced any natural object to a commodity to be bought or sold interchangeably. If you can buy centuries-old forest as development land just as you'd buy a few truckloads of cement, what's the cause for concern?

**

The twentieth century saw the greatest triumph of rationalism, uniquely symbolized by Chekhov's cherry orchard, cut down for holiday homes.

Finally, assessed of his own power, man convinced himself that he was able to solve any problem through rational organization and improved technology. "Trusting in science as a new religion, he decided to establish his rule and direct the development of nature and society according to his whim," said ecologist J. Blanco (1995). How couldn't he trust in it? A hundred years passed between the first light bulb (1879) and the first mass-produced personal computer (1981). In that time, man had invented the internal combustion engine and the television set, conducted a nuclear chain reaction and delved into the secrets of genetic code. Humanity had covered the Earth in a web of high-speed railroads and highways, and wrapped it in a thick but invisible cloud of electromagnetic beams identifiable even from space. We united three oceans by way of canals, dammed up the world's great rivers and brought together islands and continents with bridges and tunnels.

In one century, humanity brought to life all of its most audacious dreams from the "water of life" (antibiotics, organ transplants) to the flying carpet airplane, and from the Tower of Babel to Jules Verne style moonshots. By the way, that last one, the Apollo Program first thought up in the Kennedy administration, less resembled in its methodical planning a hubristic challenge to the gods and more the routine launch of a new aircraft carrier. So punctually, stage by stage, without a single misfire was that fantastic and insanely expensive ($25 billion) project realized, right up to the landing by first earthlings Neil Armstrong and Michael Collins upon the moon's surface[1].

What could stop this triumphal march of scientific progress, which had opened before humanity new horizons of life and previously unthinkable comforts, satisfied our most fickle desires and, most importantly, provided a sense of permanence to our existence on earth, tucked snugly in our reliable cocoon of civilization? In Bunin's tale, *The Gentleman from San Francisco,* the Devil himself looks sheepishly at the "Atlantida" calmly plowing through the stormy winter sea, a fabulous

---

[1] A misfire occurred, though after the successful Apollo 11 and Apollo 12 missions, when, due to an explosion in the command module of Apollo 13, it was forced to return home without completing a lunar landing. Luckily, the return flight was successful.

multi-decked liner full of celebrating crowds in torch-lit salons, reliably insulated by deep holds from the body of a dead American returning home after his ill-starred voyage, symbolizing the horrifying triumph of self-assured technical progress.

"But there, on the vast steamer, in its lighted halls shining with brilliance and marble, a noisy dancing party was going on, as usual. On the second and the third night there was again a ball—this time in mid-ocean, during a furious storm sweeping over the ocean, which roared like a funeral mass and rolled up mountainous seas fringed with mourning silvery foam. The Devil, who from the rocks of Gibraltar, the stony gateway of two worlds, watched the ship vanish into night and storm, could hardly distinguish from behind the snow the innumerable fiery eyes of the ship. The Devil was as huge as a cliff, but the ship was even bigger, a many-storied, many-stacked giant, created by the arrogance of the New Man with the old heart. The blizzard battered the ship's rigging and its broad-necked stacks, whitened with snow, but it remained firm, majestic—and terrible. On its uppermost deck, amidst a snowy whirlwind there loomed up in loneliness the cozy, dimly lighted cabin, where, only half awake, the vessel's ponderous pilot reigned over its entire mass, bearing the semblance of a pagan idol. He heard the wailing moans and the furious screeching of the siren, choked by the storm, but the nearness of that which was behind the wall and which in the last account was incomprehensible to him, removed his fears. He was reassured by the thought of the large, armored cabin, which now and then was filled with mysterious rumbling sounds and with the dry creaking of blue fires, flaring up and exploding around a man with a metallic headpiece, who was eagerly catching the indistinct voices of the vessels that hailed him, hundreds of miles away. At the very bottom, in the under-water womb of the 'Atlantis,' the huge masses of tanks and various other machines, their steel parts shining dully, wheezed with steam and oozed hot water and oil; here was the gigantic kitchen, heated by hellish furnaces, where the motion of the vessel was being generated; here seethed those forces terrible in their concentration which were transmitted to the keel of the vessel, and into that endless round tunnel, which was lighted by electricity, and looked like a gigantic cannon barrel, where slowly, with a punctuality and certainty that crushes the human soul, a colossal shaft was revolving in its oily nest, like a living monster stretching in its lair. As for the middle part of the 'Atlantis,' its warm, luxurious cabins, dining-rooms, and halls, they radiated light and joy, were astir with a chattering smartly-dressed crowd, were filled with the fragrance of fresh flowers, and resounded with a string orchestra (Bunin 1918: pp. 56–7)."

But it is not only the elements over the rail that the author juxtaposes against the luxury, the idleness, the dancing to the string orchestra, reliably shielded by the Atlantida's glass and steel from the wintry ocean. The hellish furnaces in the ship's underbelly, "where monstrous furnaces yawned with red-hot open jaws" fed by the brute efforts of other people, "purple with the reflected flames, bathed in their own dirty, acid sweat." It is their backbreaking labor which supports the indefatigable "bouquet of life" in the dance halls and salons of the upper decks. That was the harsh reality of life at the beginning of the past century as Ivan Bunin saw it.

By the middle of that century, however, the technological progress once born of this inhuman type of labor reduced it to nothingness. Most of the coal-based fleets were sent to the scrapyard, made obsolete and replaced by more efficient and ergonomic vessels that ran on residual fuel oil. Such was also the case with child labor in the textile industry, described by Jack London in "The Apostate." They made it, and then, with the transition to a shuttle-less loom, they saw it out.

Thus, is that all we need, time and patience, in order to wait out the fruits of the common welfare that scientific progress, riding upon a wave of insatiable inventive thought, will inevitably deliver to humanity? What it has delivered, however, lies within our view. The price that it has taken in exchange does not always make itself apparent to the untrained eye.

Ilya Ilf once melancholically noted, "Here they've invented the radio, but there's still no happiness." This was in the early years of the Soviet Union, when an entire generation was living with the exhilarating dream of the impending dawn of communism. This faith in the limitless possibilities of civilization differed little in essence from that of many Western intellectuals, also filled to bursting with their own manner of social optimism.

This unshakable faith in a bright future, the psychology of victorious man to which the whole natural world would ultimately bow, broke through even onto the pages of popular science. Here, for example, is how English astrophysicist James Jeans finished his seemingly non-ideological work, *The Universe Around Us:*

"As inhabitants of the Earth, we are living at the very beginning of time. We have come into being in the fresh glory of the dawn, and a day of almost unthinkable length stretches before us with unimaginable opportunities for accomplishment. Our descendants of far-off ages […] will see our present age as the misty morning of the world's history; our contemporaries of to-day will appear as dim heroic figures who fought their way through jungles of ignorance, error and superstition to discover the truth, to learn how to harness the forces of nature, and to make a world worthy for mankind to live in" (Jeans 1929).

And nearly the same triumphal attitude, only mixed with a haughty condescension toward "base" nature (do people really think like this?) looks up at us from the pages of a different work, published around the same time as Jeans'. Thousands of miles away, in Moscow, Ilya Ilf and Yevgeny Petrov wrote this in their novel, *The Golden Calf:*

"Perhaps Russian émigrés, driven to distraction by selling newspapers on the asphalt fields of Paris, recall the country roads of Russia with all the charming detail of their native landscape: the crescent moon lying in a puddle, the crickets praying loudly, and the clanking of empty buckets tied to peasant carts. But nowadays the light of the moon has another function in Russia. It will soon shine just as well on tarred roads. Motor car horns and klaxons will replace the symphonic music of the peasant bucket. And you will be able to hear the crickets in special sanctuaries." (Ilf and Petrov 1962).

Spiritual intoxication, however, shows itself no less fraught than the physical kind: both deprive us of the ability to observe our surroundings as they are. And the reality, meanwhile, demonstrates something entirely opposite.

The two bloodiest wars in history, which, by the way, in no way delayed using all the fruits of scientific progress, unfolded in the twentieth century. Likewise, the record-breaking cruelty of totalitarian regimes—from Hitler to Stalin and from Mao Zedong to Pol Pot—also became an indelible legacy, practically the calling card of that century. It witnessed the genocide of entire nations, and an endless string of interethnic and civil wars that carried off tens of millions of lives. The most cynical criminality—from the Italian Mafiosi of the 1950s and '60s to the Russian counterparts in the '90s. The invisible outbreak of international terrorism that came in the final third. All of this is inseparably linked to the twentieth century. And now the gruesome torch is passed on to the twenty-first.

Cruelty, of course, even in its most extreme forms, was a part of humanity from the present back to prehistory. But then, at least, it could find explanation in those dark recesses of the soul in which our distant ancestors dwelt, lacking a true assessment of what others' suffering and death meant, or that people of a different tribe or race might think and feel like oneself. But how do we explain this string of crime and murder in broad daylight, under the watchful eye of civilization? Perhaps civilization itself set the stage for the erosion of moral and ethical norms, shaking age-old limits and taboos, which people once accepted without question as incontrovertible law?

In reality, tricks of rational consciousness in the form of various ideological clichés, the conception of a "true" faith, of the "correct" worldview, of the "existential interests" of one's clan or social group allow a person to violate any and all primeval injunctions with an easy conscience whenever they touch upon representatives of a different nation or social caste. Thus it was, for example, with Hitlerism or Stalin's tyranny.

But even more dangerous is the utilitarian consumer attitude toward the natural environment, which dictates the rational consumer consciousness of modern man. The character Yevgeny Bazarov, from Turgenyev's *Fathers and Sons*, puts this sharply and cynically: "Nature is not a temple, but a workshop, and man plies his craft in it." He plies his craft, we might add, thinking only of today, and does not stop even when the workshop is collapsing.

By the way, unlike Bazarov, the best minds of the age understood this perfectly well at a time when neither the term nor concept of ecology existed. In one scene of Chekhov's *Uncle Vanya*, Doctor Astrov tries to explain to prominent St. Petersburg socialite Elena Antryevna the situation in one of pre-revolutionary Russia's distant counties with the aid of a homemade topographical map.

"Look there! That is a map of our country as it was 50 years ago. The green tints, both dark and light, represent forests. Half the map, as you see, is covered with it. Where the green is striped with red the forests were inhabited by elk and wild goats. Here on this lake, lived great flocks of swans and geese and ducks; as the old men say, there was a power of birds of every kind. Now they have vanished like a cloud. Beside the hamlets and villages, you see, I have dotted down here and there the various settlements, farms, hermit's caves, and water-mills. […] Now, look lower down. This is the country as it was 25 years ago. Only a third of the map is green now with forests. There are no goats left and no elk. The blue paint is lighter, and so on, and so on. Now we come to the third part; our country as it appears to-day. We still see spots of green, but not much. The elk, the swans, the black-cock have disappeared. It is, on the whole, the picture of a regular and slow decline which it will evidently only take about 10 or 15 more years to complete" (Chekhov 1999 [see URL in bibliography]).

And neither Chekhov nor his protagonist makes any secret of the direct link between the deplorable state of surrounding nature and the spiritual world of those living alongside it. He makes this far from medical diagnosis: "…It is the consequence of the ignorance and unconsciousness of starving, shivering, sick humanity that, to save its children, instinctively snatches at everything that can warm it and still its hunger. So it destroys everything it can lay its hands on, without a thought for the morrow …" (Chekhov 1999).

But could it be that industrially developed states, with the better-equipped and more prosperous lives that Doctor Astrov dreams of, a different fate awaits preserved nature? Not by a long shot, as well illustrated by the example of Western Europe and the USA in the late nineteenth century, where the chopping down of forests, the tilling up of prairies and the extirpation of wild animals and birds took on an even greater scale than that of relatively backward Russia. The oncoming twentieth century would lead these countries to the edge of total ecosystem annihilation, transforming both regions into global zones of environmental destabilization.

As the famous Russian actor-director Rolan Bykov said in his final televised interview, "Nature first died in people's souls and minds, as a focus of their energy, when nature wasn't important." There could be no better words to reveal the deeper cause of the current ecological crisis.

Sure, you can't say that modern man is totally apathetic to its beauty—he hasn't turned into a mechanical robot. Wherever he has the chance, he's not against finding himself a picturesque nook and will even spare no expense in arranging the greatest conveniences there. You need only look to how the nouveau-riche of every stripe have settled around Russian national parks, as is well known.

But that's personal. Where it concerns calculated business decisions, political struggles of bureaucratic careers, nature enters in as simply a means, the ends of which each actor chooses for himself. As a rule, they choose the end most relevant to them today: the commercial success of their firm, victory in the next election cycle, achieving the strategic military advantage of their country, and so on. It's not hard to tell that behind each aim stands, as a rule, short or at best, medium-term interests. Naturally experienced more acutely, they squeeze out of people's consciousness long-term interests connected to preserving the environment and the fate of future generations.

In essence, this is a no-man's land, territory lacking a truly concerned steward. What is perhaps the most grandiose drama in the history of life on Earth is unfolding on this territory, though most people, especially city-dwellers, know it only second hand—from the news or televised features—and thus do not take it to heart. At its base is the long-running conflict between Man and Nature in which the latter, step by step, suffers a crushing defeat. In earlier times, things stood the other way around, with the role of the conquered played more often by man. At some point, the pendulum swung to the opposite side, and now Man himself, by right of conquest, dictates his conditions to nature, while the other, one by one, surrenders its positions.

Granted, it is a temporary, pyrrhic, victory though the biosphere is suffering many irreplaceable losses in the bout. Nonetheless, in its nearly four billion years of existence, surviving more than one global environmental transformation (such as the conversion from a reducing atmosphere to an oxidizing one), the biosphere has always found ways to carry on, resetting genetic programs and cutting off the development path of those species that destabilized it at a given stage. It's entirely possible that in the distant past, large dinosaurs played such a role, and life cut off that dead-end evolutionary path.

But now these kinds of "defense" mechanisms have been activated, apparently, against humans themselves, a far more powerful disruptor of the environment than Mesozoic lizards. A growing number of genetic and congenital abnormalities, a lowered immune status and reproductive function, a mass predisposition toward diabetes, epidemic allergen sensitivity, a wave of psychiatric illnesses, especially in developed countries—all of this is an obvious consequence of deepening conflict of "human nature" with the technologized conditions of its existence. Furthermore, humans, like the rest of the global biota, must carry the burdensome pernicious influence of the toxins and allergens, carcinogens and mutagens that enter the biosphere in the process of their economic activity.

No less serious a threat is presented by the human-caused reconstitution of processes in the viral, bacterial and fungal communities that accompany us, provoked by invisible interspecies wars both hot and cold. Likely confirmation of this comes in the form of several viral infections that transfer from the animals we have supplanted, namely, AIDS, bird flu, atypical pneumonia and Ebola.

In this we can see the apparent action of negative or compensatory feedback, directed at suppressing man as the source of biospheric disruption. That is, the biosphere destroys its destroyer. Granted, we still have the strength to oppose this harmful trend, drawing upon our scientific and technological power. But we must recognize that the civilized world of our time more closely resembles fortress under siege, barely managing to deflect the impacts that come in from every side.

This concerns not only the sporadic outbreaks of dangerous infections that redirect great quantities of strength and coin, or the tumult of natural disasters, the more frequent character of which has already been directly connected to human economic activity. It also concerns the "epidemic" of accidents involving human technology and their frequent ecological consequences, terrorism, interethnic warfare and other such phenomena of mounting intolerance, putting the brakes on the development of civilization from within.[2] But most tragically, we are accustoming ourselves to this state of permanent siege in which humanity will have to live, by all appearances, for an indefinitely long period of time. To see the truth in these words, it is enough to turn on any edition of television news, which more often than not is filled to the brim with this bleak chronicle.

To complete this first section of the book, which has focused on the many faceted incarnations of the global crisis, let us use just seven words: a sick biosphere and a sick society. In this formulation, however, there is no symmetry. Civilization, after

---

[2] Here, for example, is how one contemporary anthropologist looks upon the issue of suicide terrorism: "On its face, it is a previously unheard of critical-catastrophic trend in anthropology. This new frontier is not simply new, it is the last frontier. Whenever crises occurred in history—downfall and defeat, famine and pestilential plague—for centuries there existed the commonplace view that we could get over such crises with a surplus of what you might call anthropological endurance <...>. The new anthropological situation says, however, that the surplus of anthropological endurance is exhausted" (Khoruzhy 2002).

all, is a subsystem of the biosphere and not the other way around. And since, as we have attempted to show, civilization contains the primary source of this "sickness," the mechanisms of the whole will be directed against this part. This means there will obviously be blowback working against civilization for as long as it takes for the "sickness" to be neutralized. If, of course, humanity does not awaken and put an end to this mad war against nature.

References

Blanco, J. A. (1995). The Third Millennium. *Latinskaya Amerika*, (9), 4–14 (in Russian).

Bunin, I. (1918). *The gentleman from San Francisco* (A. Yarmolinsky, Trans.). Boston: The Stratford Company.

Chekhov, A. (1999). *Uncle Vanya* (Marian Fell, Trans.). Project Gutenberg eBook #1756. Retrieved from http://www.gutenberg.org/dirs/etext99/vanya10h.htm.

Ilf, I., & Petrov, E. (1962). *The golden calf* (J. H. C. Richardson, Trans.). New York: Random House.

Jeans, J. (1929). *The universe around us*. Cambridge: Cambridge University Press.

Khoruzhy, S. S. (2002). A report and discussion as part of the conference "anthropological templates of the 20th century: *L.S. Vygotsky—P.A. Florentsky: The dialogue that isn't happening*." Moscow. Retrieved from http://synergia-isa.ru/lib/download/lib/003_Horuzhy_Vygodsky_Florenski.doc (in Russian)

Krasilov, V. A. (1992). *Environmental protection: Principles, problems, priorities*. Moscow: Institut okhrany prirody i zapodovednogo dela (in Russian).

Part III
World Society: Politicians and Scientists in Search of an Answer

Chapter 7
First Steps by the UN and Club of Rome. The Computer Model That Rocked the World

The initial symptoms or precursors to the current Global Crisis first made themselves known at the beginning of the twentieth century. Some of the greatest minds of the age, such as Nikolai Berdyayev, Oswald Spengler, Jose Ortega y Gasset, Erich Fromm and others, repeatedly tried to attract attention to various aspects of the problem. But two cruel world wars that crushed the whole first half of the twentieth century beneath them, followed by the exhausting standoff between two social systems (the so-called Cold War), squeezed all seemingly less important issues from humanity's field of vision.

The economic boom that followed the Second World War and the accompanying arms race with its mindless expenditure of resources brought about a new, even more sharply escalating extirpation of nature which now included the former colonies, the "third world" countries. Furthermore, the postwar decades were marked by a series of atomic weapons tests in the atmosphere and the feverish development of the chemical industry, including the production of polymer materials and the introduction of chemical fertilizer and pesticides into agriculture. It should then come as no surprise that, from the early 1960s, the global public began to feel troubled by the issue of environmental pollution and the preservation of nature.

In 1961, the Economic and Social Council of the UN (ECOSOC) adopted the historic resolution number 810, which brought up the need to create a worldwide chain of nature reserves and specially protected areas. That year also marked another important event in the environmental sphere with the creation of the World Wildlife Fund (WWF). The fund began by financing environmental protection works in the Galapagos Islands, that unique natural landmark, and went on to make a major contribution to preserving many endangered species. Five years later, in 1966, the International Union for the Conservation of Nature (IUCN) released the first international Red List, containing an extensive catalogue of world flora and fauna species under threat of extinction. This formed the basis for systematized work in the field, and similar books containing lists of rare and disappearing species in various world regions were thereafter published in many countries.

© Springer International Publishing AG, part of Springer Nature 2018
V. I. Danilov-Danil'yan, I. E. Reyf, *The Biosphere and Civilization: In the Throes of a Global Crisis*, https://doi.org/10.1007/978-3-319-67193-2_7

During the same years, another very important measure was considered and brought into being under the aegis of UNESCO, the International Biological Program (IBP). Its realization took 10 years, from 1964 to 1974, and made an enormous contribution to studying the structure and operating principles of various ecosystems. IBP participants organized numerous land and sea expeditions enabling the creation of a kind of inventory of preserved natural resources and an overall evaluation of the state of the Earth's biosphere. UNESCO's General Conference confirmed the first results of the research in 1970 with the adoption of the international Man and the Biosphere Programme. This program was invoked to bring the problem of biospheric sustainability under human pressure to the attention of the scientific community worldwide.

And these first serious steps toward clarifying the global ecological situation and the outlook for development connected to it did not limit themselves to the UN. No less a contribution to the understanding of this problem was made by the Club of Rome. An international nongovernmental organization uniting political leaders, businessmen and scientists from different countries of about a hundred members, their concern for the fate of the world would not allow them to throw up their hands at the symptoms of the unfolding crisis, but roused them to look for new development paths. Aurelio Peccei (1908–1984), Vice-President of Olivetti, took the initiative in the club's founding. The Club of Rome took its name from the Italian capital, where its founding members held their first meeting at Accademia dei Lincei.

Once it had recruited leading specialists in the area of system prognostication, the organization's main task was to research the crisis issues in all their complexity, incorporating all its aspects and related disciplines. The research projects that Peccei initiated were funded by a number of major companies and touched on various sides of the planet's crisis state. The international team of scientists who carried out the work then put it into the form of Reports to the Club of Rome.

At the time, futurology was more actively seeking to pose itself on exact mathematical methods. By no coincidence, the club recruited MIT Professor Jay Forrester, the first to use computer modeling in the study of complex tendencies of global development, including its demographic, agricultural, industrial and resource aspects as well as environmental pollution (the World1 and World2 Models). Forrester, however, handed off the new assignment to his pupil, 26-year-old assistant at MIT's System Dynamics Group, Dennis Meadows. The small team of young enthusiasts that he put together, including his wife, Donella Meadows, managed to calculate a number of scenarios for global development for the period from 1970 to 2100 on the basis of the improved World3 computer model.

According to model dynamics, at current rates of resource expenditure, industrial and population growth, human civilization, forced to direct more and more capital toward support for the environment, should hit critical barriers in its development sometime in the mid-twenty-first century, replete with economic and demographic collapse. In 1972, this work, signed by Donella and Dennis Meadows, Jorgen Randers and William Behrens, was presented and published as the First Report to the Club of Rome under the name *The Limits to Growth* (Meadows et al. 1974).

Of course, any computer model simplifies and schematizes reality to one extent or another, particularly when global processes serve as the object of the modeling. However, the impossibility of experimenting on the real object makes it an irreplaceable tool, allowing the researcher to play out different scenarios and watch to what result one or another set of parameters in the program leads. The Meadows team turned World3 into just such a tool. It was significantly deeper and more complicated than the few models then in existence used to evaluate the longer-term prospects for global development.

Included among the key parameters in the World3 model were Earth population, food production, volume of industrial production, reserves of unrenewable resources, level of environmental pollution and anthropogenic burden (humanity's ecological footprint) as well as a set of social parameters, among them goods and services per person and average life expectancy. And these were only the most important characteristics shown on the graph. There are actually many more.

By the standards of the early 1970s, World3 was equipped with a large number of cross-connections and feedback loops which allowed it to calculate their multifaceted influence on the researched processes. For example, when consequences begin to affect their own cause, as often happens in real life. Another unique aspect of the model was non-linear connections. There would be disproportionate increases of decreases in one of the parameters in response to the change in another, which take on particular meaning in finite situations when the limits come at you unexpectedly and unmanaged problems explode like a volcano. Such non-linear thinking and feedback loops combined with a systems approach to the research subject, looking at the environment, population and economy in their dynamic unity created the conditions for a life-like model. After all, these kinds of system effects, without exaggeration, penetrate our lives at every level.

Along with this, World3 knew neither war nor corruption, not crime or terrorism, and the people in this computerized world decided global problems without a thought of political advantage, ethnic intolerance or national egotism. Thus, reality here was presented in an admittedly simplified and idealized way. However, the authors guessed these simplifications, in principle, did not change the picture so far as the aim of the research was to uncover not quantified characteristics but tendencies, trends and system behaviors. How would it respond to one or another level of environmental pollution? How would the exhaustion of unrenewable resources pan out? What possibilities were there for population growth if you calculated the finite capacity of the biosphere, etc.

The model was used to run ten basic development scenarios. Each of them represented a particular type of global system behavior depending on conditions given at the outset. For example, if reserves of unrenewable resources turn out to be significantly greater than specialists believe (which is theoretically not out of the question at all), or if society decides to reorient the economy toward resource-saving technology and concentrates its investments in that area. All scenarios were characterized by corresponding numerical value inputs, which were entered into the model for a computer run. For each of the given scenarios, World3 calculated the governing parameters (over 200) and produced values for all variables for each half year

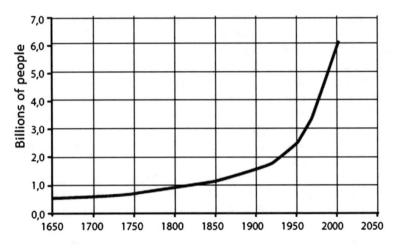

Fig. 7.1 World population from 1650. Source: UN

up to 2100. In total, for each of the scenarios, the model gave 80 thousand numerical values, which in the early 1970s was nearly the limit of computational abilities (Meadows et al. 2006).

Of course, all this would hardly shock a modern computer modelling specialist, but then, in the early 1970s, such experiments in mathematical models came as a novelty even to experienced professionals. And as we've said, the authors of *The Limits to Growth* had focused their attention primarily upon the main behavior particularities of the system relevant to the most diverse scenarios. The first of these, *exponential growth,* served as one of the chief causes of going beyond the limit. It differs from linear growth, associated with direct, uniform progression, in that the rate of growth increases simultaneously with that of each number, metric, volume, etc. As a result, at a certain moment, growth becomes explosive. The classic example is the reproduction of a yeast colony, whose cell numbers double every 10 min. Under ideal conditions, the colony would envelop the entire Earth in a relatively short period of time.

A no less illustrative example would be the colonization of Australia by a warren of rabbits, brought there by an English farmer for sport hunting in 1859. At first there was just a dozen. But on finding suitable conditions—an abundance of food, an absence of control species (predators and parasites)—the rabbits began furiously multiplying, and the population increased to 22 million in just 6 years. By 1930, the rabbits had settled the entire continent, and their numbers had reached 750 million! As a result, food supplies for flocks of sheep sharply declined, which seriously compromised the country's most important branch of agriculture. Sheep numbers were cut in half. Worse still, the rabbits were taking food from kangaroos and other herbivorous marsupials. They only managed to solve the problem in the early 1950s, when the uninvited occupants were deliberately infected with myxoma virus and their population fell by 90% (Mayr 1970).

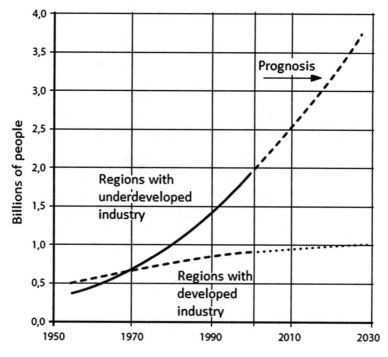

Fig. 7.2 World urban population (data from 2000). Source: UN

The growth of the human population on Earth has also been exponential for the last three and a half centuries (Fig. 7.1). In 1650, on the eve of the Industrial Revolution, there were about half a billion people, increasing yearly at a rate of 0.3%, corresponding to a doubling time of 240 years. By 1900, 1.6 billion people lived on Earth, the growth rate having increased to 0.7–0.8% a year, equivalent to a doubling time of 100 years. By 1965, the number of people on Earth already added up to 3.3 billion and growth rates had risen to 2% a year, so doubling time had decreased to 36 years. Truly, in 1999 those numbers, in total correspondence with projections, reached the six-billion mark. And while yearly growth rates started to decline in the 1980s, the increase in absolute numbers continues to this day.

Today the decisive contribution to this hyperbolic growth is made by economically backward countries. As you can see from the graph of population growth in cities in the second half of the twentieth century (Fig. 7.2), the average doubling time of urban populations in countries with underdeveloped economies is 19 years. One supposes that this trend will continue into the coming decades.

But not only populations of living beings—humans, yeast or rabbits—tend toward exponential growth. Industrial and financial capital also grows exponentially. Factories produce steel, cement, materials and equipment, machine tools and conveyor belts. A certain share of production is used to reinvest in this very capital, creating a base for future production growth. You could say we're dealing with the "birth rate" of capital, which results in the creation not only of new plant, but of new

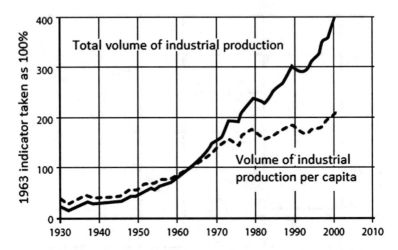

Fig. 7.3 Global industrial production. Annualized growth rates for the last quarter of the twentieth century were 2.9%, which corresponds to a 25-year doubling time. Production growth per capita was 1.3% a year, equivalent to a doubling time of 55 years. Source: UN

factories. Any commercial activity is also directed at taking in profit, which in turn will be invested in the expansion of the commercial activity and a new increase in profit (Fig. 7.3).

In cybernetics this type of phenomenon is called a positive feedback loop—when a change in some element in the system leads to a chain of results enabling still greater change in the input elements in the same direction. Increases lead to still greater increases, while decreases lead to ongoing decreases. Thus, for example, a rise in average surface temperature as the result of carbon dioxide emissions into the atmosphere speeds the melting of the permafrost zone. The melting tundra releases trapped methane, a strong greenhouse gas, which provokes further increases in the global temperature and so on. Which is to say that a process once launched will feed and stimulate itself.[1]

Of course, the realization of exponential growth requires the corresponding circumstances, and any number of factors can put a stop to it, from a lack of nutrient matter in the case of yeast to social shockwaves, war, hunger and epidemic when speaking of humans. But if one or another parameter is set to create a positive feedback loop, that means it is potentially subject to exponential growth. The input effects are strengthened by the system itself. This is the basic mechanism of the demographic explosion. Against the backdrop of a weakened stabilizing influence of negative feedback (decreasing mortality in developing countries), we as absolute

---

[1] In nature and technology the majority of regulatory processes are based on the principle of negative feedback, which supports the stability of the given system. In essence, any divergence of a system from equilibrium initiates such events that would slow the changing of its characteristics, increasing as the system goes further from its equilibrium state.

masters of fate have created a completely unlimited positive feedback loop, radically transforming the demographic situation on the planet.

In modern society, the main generators of exponential growth are population and industrial capital. Other factors—food production, resource use, pollution—also tend toward exponential growth, not because they reproduce themselves, but because they are stipulated by population and capital growth. Demographic and industrial growth bring ever greater quantities of energy and raw material into the economy as a secondary characteristic. The World3 Model expressed these structural peculiarities of the global system.

Another important peculiarity of large dynamic systems, including the global ecology and economy, is the *lag factor*. An ocean liner moving at the speed of 22 knots cannot suddenly change course once it sights an obstacle ahead. The more time it takes the ship to turn, the further its radar ought to see. The liner by the name of human civilization has incomparably greater force of inertia, and, therefore, the signals people receive of encroaching limits should not only be perceived in a timely manner, but properly interpreted.

In their book, *Limits to Growth: The 30-Year Update* (which we will discuss later), the Meadowses and Randers take a lesson from the "Ozone Story," a fairy tale with a happy ending. The story begins in 1974, with a publication by American chemists Frank Sherwood Rowland and Mario Molina (Nobel Laureates in Chemistry, 1995) warning that the release of chlorofluorocarbons (CFC) into the atmosphere threatened to destroy the Earth's Ozone Layer. These seemingly inoffensive chemicals were used in refrigerators, aerosol spray cans, pharmaceutical production and other things. Told in all its details, the story has all the makings of a pot-boiling novel, but for now, we turn our attention only to the lag point, particularly demonstrative when things concern the reaching of a critical limit. Here the discussion was of nothing more or less than the destruction of the ozone shield of the planet, which defends all living things from the harmful effects of harsh ultraviolet cosmic radiation.

From the above-mentioned first publication, e.g. the first proper human assessment of the distress signal, to the 1987 signing of the Montreal Protocol, which placed progressive restrictions on the wide-scale use of CFC, a whole 13 years went by. Then came nearly another 2 years until the protocol went into effect on January 1, 1989. Scientists spent a lot of energy convincing politicians and businessmen of the reality of the threat hanging over humanity. After all, there was big money at stake, and businesses would have to undergo great expense in order to find a suitable ozone-safe replacement for CFC and refit the factories that produced them. And while it would have seemed that scientists' conclusions had confirmed the broadening hole in the ozone above Antarctica, new, plausible-sounding explanations for the phenomenon were then thought up. Many companies, interest groups and entire countries took umbrage, declaring that they would defend their corporate and narrow national interests at any price. It required another 13 years to achieve total implementation of the Montreal Protocol to which, after numerous amendments and revisions, 157 governments agreed. After that, the ban on the production and use of substances harmful to the Ozone Layer took on force of law.

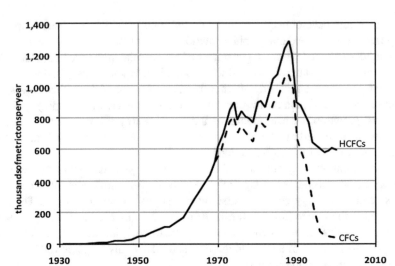

Fig. 7.4 Worldwide chlorofluorocarbon (CFC) production. Source: (Meadows et al. 2006: p. 183)

But the 26 years that passed after the harmful effects of CFC to the Ozone layer were found was just the initial lag stage, following the typical pattern of human activity according to the "discovery—implementation" Scheme. A second lag stage, already beyond the boundary of technology, arises from the reaction of nature to human action. This one is lengthier, since CFC molecules, distinguished by their persistency, leave the upper levels of the atmosphere very slowly. The thing is, it takes CFC molecules ten-twenty years to reach the Ozone Layer. At the same time, one CFC molecule can become a catalyst for several thousand reactions that destroy ozone molecules by turning them to common oxygen ($2O_3 \rightarrow 3O_2$) before it reacts with other substances (methane, for example), which can take many decades. Thus, for example, Freon-12, one of the most widely-used CFC's and whose production came to an end only in 2010, persists in the atmosphere for about one hundred years, and equipment full of it is still in service. So plenty of time will pass before ozone-destroying substances, once they have left junked refrigerators and cooling units, are carried up to the ozone layer and then begin producing their pernicious resource. To return concentrations of ozone to 1980 levels, we may have to wait until no earlier than the mid-twenty-first century.

After the Montreal Protocol went into effect, world-wide CFC production fell from one million metric tons (1.1 million standard tons) in 1988 to 100,000 (110,000) tons per year by 2000. However, ozone concentrations continued to fall and the ozone hole over Antarctica continued to expand until 1997, after which concentrations started to slowly increase. In 2007, in an amendment to the Montreal Protocol, parties adopted a decision to speed the phasing out of a group of the less harmful hydrochlorofluorocarbons (HCFCs). According to the agreed stipulations, all developed countries were obligated to reduce HCFC production and use by 90%

before 2015. (website for UNIDO/GEF-Nature Ministry of Russia http://www.ozoneprogram.ru/ozon_sloi/sohranenie_ozona/).

The trends in worldwide CFC production shown in the graph (Fig. 7.4) serve as a clear illustration of what we have said. The years from 1950 to 1974 are marked by sharp growth in CFC production, followed by a short-term fall as the result of "green" activism sounding the alarm concerning messages from scientists and the expanding ozone hole over the poles. However, the effect of social intervention, as you can see from the graph, was quite short-lived and quickly replaced by charging growth in CFC and HCFC production brought about by an expanded field of applications. Only in the late 1980s, after the Montreal Protocol went into force, did the irreversible decline in chlorofluorocarbon production begin. This was more distinct in the case of CFCs and not quite as expressed with HCFCs, which were still permitted for use in the early 2000s. The 2007 adoption of the amendment to wind down their production, however, allows us to hope for a complete end to its sale in the coming years (Meadows et al. 2006).

In this way, the lag factor is a fundamental quality of all complex systems such as the biosphere and civilization. Between the release of a pollutant into the environment and the moment when that begins to tell upon the health of a population, a certain amount of time will pass, sometimes a long one. Decades are required for the redistribution of investments compelled by food shortages and soil degradation. For people to transition to two or three children from the traditional large families as child mortality declines, one or two generations must pass. Even if a system reacts in time to the danger and distress signals it receives, it cannot change itself overnight.

Thus, in the population of countries undergoing a demographic boom, there is a very high share of young people. Therefore, however successful measures to control the birthrate may be, the population will still continue to grow for at least several decades—until the numerous youth born during demographic boom times passes its childbearing years. And while the quantity of children in the average family will contract, the overall number of families will increase. Such is demographic inertia, which will not allow us to stop population growth tomorrow or the day after. And if by some miracle we managed to reduce births to the bare replacement rate all at once, all the same, we would still have to wait several decades for the Earth's population to stabilize.

***

The Limits to Growth, coming out in the early 1970s, stirred an unheard of public response. The book was translated into 35 languages and immediately became a best seller. The vast majority of readers up to then had never given thought to any "limits," naively supposing that humanity was one thing, the Earth another, and that their scales were so incommensurate that no human activity was in a position to do harm.

And while the book stirred no little disagreement among specialists, it still planted the seed and, corresponding entirely with the Club of Rome's intentions, left a shocking imprint upon the mind, not only for the casual reader, either, but also for business people and captains of industry. Management Theorist Dzherman

Gvishiani noted, "Not least of all, the dark prophesies of *The Limits to Growth* forced industry to transition to materials-saving production, to develop new technologies, re-use resources, create new synthetic materials, start economization programs, etc (Gvishiani 2002).

To the authors themselves, however, the results came as no surprise. After all, despite isolated corrective measures, the overall tendency of global development remained the same. Looking back, they wrote, "When we wrote LTG we hoped that such deliberation would lead society to take corrective actions to reduce the possibilities of collapse. Collapse is not an attractive future." But such was not the case, "No modern political party has garnered broad support for such a program, certainly not among the rich and powerful, who could make room for growth among the poor by reducing their own footprints" (Meadows et al. 2006: xi, xv).

Two decades later, the same team of authors (with the exception of William Behrens) returned to the topic in the book *Beyond the Limits*. While their basic theoretical impulses remained the same, they added a note of pessimism not seen at the beginning of their careers. Such feelings were well founded. The ozone holes over the poles, global climate change, more frequent natural cataclysms, a growing freshwater shortage, wide-scale cutting-down of tropical forests, declining catches of fish at sea—all of these alarm bells and wake-up calls that up to then could still be ignored testified to the fact that humanity on many fronts had already gone beyond the limits of Earth's biological carrying capacity and was living on an area outside sustainability, e.g. beyond the limits to growth. This none-too-pleasant fact the authors put into the title of their second book.

But how irreversible is this exit beyond the limits of sustainability? Is it ever observed under natural circumstances, independent of humanity? Yes, on a local scale, it has even been observed everywhere. For example, meadow and grassland eco-systems evolve together with the herds of herbivores that graze on them. These, of course, do not worry about supporting an ecological balance, and they can strip the grassy cover bare. However, nature itself does show concern about it, and if the root system is unharmed, the remaining roots and lower stems receive more water and nourishment, causing the grass to grow again. The herd temporarily makes its way to other pastures. Thus, where the possibility of migration remains, the ecosystem is not destroyed but abides in a state of dynamic balance. And with the restoration of the extirpated plant cover, the herd can once again return to its deserted feeding grounds.

Something of this kind takes place in the sphere of human activity when its steps beyond the limits do not destroy the capabilities of the environment. Here are a few examples from the book, *Limits to Growth—The 30-Year Update*.

In the history of America's New England, there have been several instances when sawmills closed down en masse due to the exhaustion of timber supplies. The mills closed, and forestry lay in a dormant state for several decades. When the forests grew again, the buzz-saws went back to work—right up until the process of forest resource overexploitation led to the next local crisis. The coastal fisheries of Norway have gone through at least one cycle of marine resource exhaustion. The government bought up fishing trawlers and had them scrapped. Fish populations had to

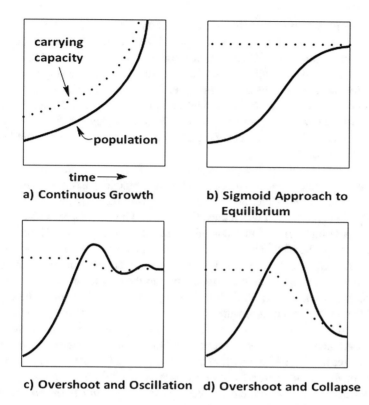

a) Continuous Growth

b) Sigmoid Approach to Equilibrium

c) Overshoot and Oscillation

d) Overshoot and Collapse

Fig. 7.5 Population growth and alternative paths of its interaction with Earth's potential carrying capacity. Source: (Meadows et al. 2006: p. 138)

recover enough to allow a return to the traditional trade (Meadows et al. 2006). The writers called this "overshoot and oscillation"—when the destruction of renewable resources is not irreversible and does not undermine the ability of life systems to restore themselves.

These examples, however, relate to instances of localized steps beyond the limits. How will human civilization as a whole behave, and what will come of exceeding the Earth's carrying capacity as the population grows? In Fig. 7.5 we have presented four graphs—the four hypothetical possibilities for the unfolding of events. The first graph (Alternative *a*) illustrates *uninterrupted (hyperbolic) growth* in the global population under circumstances when the limits are still far away, somehow do not exist or themselves grow exponentially, going off alongside the population line (which to a certain extent corresponds to the situation on Earth at the start of the previous century.)

Alternative *b*, an *S-shaped (sigmoid) curve*, illustrates growth in the population of living organisms under conditions of finite food resources and environmental resistance, when population growth slows as it approaches the limit before stopping at a state of dynamic balance. Applied to human civilization, such an alternative is possible if the system (the population and economy) consciously limits itself and reacts in a timely manner to signals of approaching limits. However, for the seven-billion-strong population of Earth, whose numbers are already beyond the limits of the Earth's potential carrying capacity, this opportunity has practically passed us by already. As Meadows and his co-authors note, "The simplest and most incontrovertible physical delays are already sufficient to eliminate smooth sigmoid as a likely behavior for the world economic system". Therefore, only the final two graphs remain relevant today—going beyond the limits with fluctuation or going beyond the limit with catastrophe.

Going beyond the limits with fluctuation ("overshoot and oscillation"). This possibility, by all appearances, is not yet lost to humanity so long as our sojourn beyond the limits remains reversible. Such was the case with the restoration of the ozone layer after CFC production was reduced. Such could it also happen with the greenhouse effect resulting from anthropogenic carbon dioxide emissions. Unfortunately, in the opinion of the Meadows team, it typically turns out differently: before taking action in the right direction, people first go beyond the limits, and only later, either on their own or facing pushback from nature, do they try to return to the zone of sustainability (Meadows et al. 2006). The kind of system behavior where going beyond the limits is not accompanied by irreversible changes to the planetary environment and can be stopped and turned back by corrective measures is shown in *Graph C*. At the same time, due to inertia and the lag in the system, the limits may be crossed repeatedly, taking on the character of waning fluctuation, as you can see.

And, finally, *going beyond the limits with catastrophe ("overshoot and collapse")*. One well-known principle of ecology is *population equilibrium*: Stability in the population of any species occurs as the result of dynamic equilibrium between its biotic potential and resistance from the environment (temperature extremes, limited food supplies, predators in the ecosystem, etc.). But, in the case of the human population, this feedback doesn't work. In creating an artificial habitat, people have provided themselves with relative independence from the planetary environment, which means that the prerequisites for totally unencumbered growth have also been provided, including a material base for the necessities of life. It's as though humans have separated themselves from the biosphere, and that has caused them to go beyond a string of limits. If anything from outside can stop it, that would be global ecological catastrophe, the likelihood of which also grows exponentially. So it should come as no surprise that most of World3's calculated scenarios ended somewhere before the end of the twenty-first century with a planet-wide crisis of renewable and unrenewable resource exhaustion, farmland erosion, shrinking food production and, consequently, a momentous collapse of the population.

In recent times, "green" internet pages have begun to fill with advice about how to minimize the harm each of us does to the environment. For example, there's "50 simple things that help save the planet": Brush your teeth with the water off, turn in used cans and bottles, buy reusable batteries instead of single-use, etc. But, as the Meadows team put it, "All these actions will help. And, of course, they are not enough…Recycling is important, but by itself will not bring about a revolution" (Meadows et al. 2006).

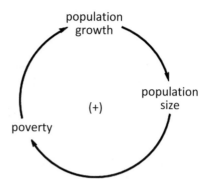

Fig. 7.6 Poverty and population. Source: (Meadows et al. 2006: p. 45)

So, what do we need to save the planet? Most of all, these are structural reforms which, in the opinion of the authors, could neutralize the very causes of this exit beyond the limits and exponential growth. For example, glaring social inequality in underdeveloped countries where, in the absence of stabilizing mechanisms to level the playing field for everyone, society's privileged classes acquire ever greater power and resources, ensuring the way is open to future riches. As a result, the rich get richer and the poor get poorer, hopelessly mired in a fen of poverty and despair which, as you know is inseparable from demographic growth.

This ends in a system trap as growth creates poverty and poverty creates growth, forcing society to remove wealth from the investment cycle for use on consumption. In short, it is eaten up. In this way, the two factors join together in a positive feedback loop, strengthening each other as illustrated in Fig. 7.6. And there is only one method to break this vicious cycle—a deliberate investment policy designed to provide access to education, health care, and family planning programs for the poorest strata of society, particularly for women. After all, in the absence of attractive alternatives to childbirth, when there is no opportunity to work or to learn, children become their lone and most important capital.

But, could you not say the same about the polarization of wealth and poverty in the world as a whole? It is easier for rich countries to save and increase their capital acquired over hundreds of years of economic development, and slow population growth enables them to invest greater means in economic expansion. Poor countries are forced to spend the lion's share of their resources on satisfying the urgent demands of a growing population, at the expense of economic and social development. Thus, only simultaneous restructuring of the consumption model in developed countries combined with the targeted use of freed-up funds in countries that badly need them would allow the untangling of the ball of problems connected to exponential population and capital growth and the fatefully growing burden on the environment.

***

As we've already said, *The Limits to Growth* met with plenty of reproach among specialists, while supporters of its approach were in short supply. Criticism sounded louder than approval, but the unheard-of success among readers roused many scholars to work on improving the World3 Model as well as alternative programs. What failed to suit these detractors?

First of all, it's worth saying that among the dissatisfied were representatives of many fields of knowledge: economists, biologists, geologists, mathematicians, sociologists, political scientists, science theorists and even philosophers. And all of them, each on behalf of their discipline, chided the authors for oversimplification. The world was not as simple as the World3 Model, they said, and brought forth examples of global phenomena and processes without which World3 was incomplete and unrepresentative. The model did not properly express progress in science and technology, for example. Although it was well known that the introduction of new technology reduced resource usage rates, making them more efficient and reducing the negative effect of industry upon the environment on a per-unit basis.

Pointing to this shortcoming, the critics spoke of the necessity, without jumping to foregone conclusions, of depicting the influence of these advances on resource prices and reduced ecological consequences on a quantitative level. Naturally, this was a far from simple task, and most critics were aware of that, knowing that many scientific and technological expectations either never come true at all or arrive with great delay. Therefore, the character of scientific advancement's influence on global development, much like the hypotheses based upon it, are not distinguished by statistical certitude. In the 1960s and '70s, for example, few people doubted that by the beginning of the twenty-first century we would have developed controlled thermonuclear fusion. Today, you'd be hard pressed to find anyone willing to make a solid prediction on that account, and many doubt the practicability of its development: high water usage by controlled thermonuclear reactors could make them uncompetitive compared with renewable sources of energy. At the same time, nobody in 1970 could have guessed that in 30 years cell phones would conquer the world, or the arrival of the internet which has revolutionized our whole lives. It's worth recalling the panic that ensued in the 1950s concerning what then looked like the impending exhaustion of silver reserves widely used for black-and-white film. Once color film took over, the problem disappeared. Then on to digital technology. The list of such examples goes on and on.

Critics also made note of the self-evidence of the math conclusion in *The Limits to Growth* concerning the basic finiteness of a resource used by a growing system whose usage volume increases as it grows (as is the case under exponential population growth on Earth.) To many mathematicians themselves, Forrester's system dynamics model at the base of World3 appeared too elementary.

Another point of criticism for the first report to the Club of Rome was the concept of "zero growth" presented within. In a 2007 lecture, Dennis Meadows quoted a newspaper column of the time as saying that a non-growing economy was hard to imagine, and could lock poorer countries into poverty (Meadows 2007). President George H. W. Bush expressed it even more categorically: "Twenty years ago some spoke of limits to growth. But today we now know that growth is the engine of change. Growth is the friend of the environment." (Quoted from (Meadows et al. 2006)).

What do we see most in that turn of phrase? Populism of the worst kind, giving preference to short-term interests over that which is "hard to imagine," as it concerns not the present but future generations (in full correspondence with the famous formula "After us—the deluge?") or the lack of desire to understand the seriousness of the threat hanging over humanity? On the other hand, one must always take into account the expansionist yearnings, native to the majority of people—to increase their possessions, their material wealth, their business, etc., — which are genuinely incompatible with the understandings "limits to growth" and "zero growth." So it also is, by the way, with the strategy of life itself, which is based on a combination of expansion and sustainability. So we should clearly show no surprise at the opposition the Meadowses and their co-authors faced concerning zero or, what's more, negative growth. But while, in the early 1970s, they had the power of novelty with which to drown out some of the opposition, later on, once the novelty had faded, human "nature" came back with a vengeance, returning to fight objective natural laws.

But we will not debate which is more correct: "growth is the engine of progress" or "progress is the engine of growth". The matter is how to understand the words *growth* and *progress*. In the context of the Meadows team, growth is understood most of all in material terms, measured in physical mass and energy. That is, the growth for which limits are set. Unfortunately, for many critics of their concept, including the 41st US President, clearly no other growth presents any interest. At the same time, the human spiritual being as well demands cultural development, a deepening of scientific knowledge, improved social relations, philosophical and religious inquiry, among other things. And for growth of this kind, provided relatively insignificant material expense, obviously no limits can exist. Thus zero growth is in no way a stop to development, but primarily a stop to the negative or harmful influence on nature.

However, given the catastrophe threatening the biosphere, a different type of question is justified: Is it worth focusing on growth as the source of the problem? After all, it is not growth itself that presents the threat, but the accompanying deformation and destruction. Concentrating our attention on the latter, also inseparable from modern civilization, would be more precise. Within the bounds of this destruction, we convert the problem to a more constructive course, allowing us to use more concrete signposts, not only in ecology but in the socio-medical and humanitarian spheres. These, too, hold no shortfall of threat to human survival. This includes the undermining of population health in the species *Homo sapiens*, the increasing destructive power of social processes and much else, without which it is impossible to speak in any way of sustainable development. And beyond that, unlike limits to growth, the idea of limits to destruction harmonizes better with human nature, which is still wont to look upon the work of its hands, including the destructive consequences of its own activity.[2]

In 2004, the same team of authors (with the exception of William Behrens) again returned to the topic of limits and released another already mentioned version of their book entitled *Limits to Growth: The 30-Year Update* with a more fully developed thesis about the basic finiteness of Earth's resource potential and a wider pan-

---

[2] For more on the concept of limits to destruction, see (Danilov-Danil'yan 2003).

orama of crisis phenomena in the modern world. A lot of time had passed since the first publication, which allowed not only for the integration of rich and accumulating information-science resources, but also for the comparison of global development's trajectory with the variations observed in 1972. Surprisingly, an exact correlation of indicators for real development was found with the inertia scenario from the early 1970s.

What does that tell us? Most of all, that humanity has not come to take the serious measures to ensure stability in global development. Even today, 40 years after the initial publication, worldwide assessments of the environment continue getting worse. The only hopeful sign on this horizon was the production of ozone-destroying substances and the increased concentration of ozone in the atmosphere. It was for this reason that the 1972 inertia scenario turned out so close to reality. Therefore, the nature of scientific and technological innovation also corresponded to inertial development, and the exceptions—renewable energy production and energy-saving methods—did not change the overall picture one bit, being entirely insufficient to right the ship of civilization. That, perhaps, is the main lesson that humanity can take from the retrospective evaluation of the first report to the Club of Rome. But, my-oh-my, we haven't taken it.

For the sake of fairness, it is worth recalling other, alternative models for global development designed after the pioneering research by the Meadows team (1974). In part, in the second report to the Club of Rome, *Mankind at the Turning Point*, Mihajlo Mesarovic and Eduard Pestel put forward the idea of "organic growth", according to which the various world regions coordinated harmoniously with one another would develop each in accordance with its own specifics while remaining in concert with the interests of the whole analogous to the development of a living organism (Pestel 1989). However, the modeling methodology of the second report turned out to be even more tenuous than the first. The accumulation of large volumes of information about the world's regions and the complication of the model did not lead to materially new results. More importantly, neither this nor any following report to the Club of Rome caused the type of resonance that *The Limits to Growth* brought about in the early '70s, like the striking of an alarm bell, like a warning to a humanity intoxicated with the success of civilization about where these successes would lead.

Today the World3 model and the books composed by its creators are studied at many of the world's universities. These works unquestionably influence active members of the older generation as well—politicians, business people, scientists—whose decisions in many ways determine the future of our planet. The authors, though, harbor no illusions on this account. After all, the discussion concerns restructuring the consciousness of entire nations, changing the systems of values, guide posts and stimuli for life that force people to spend natural resources more prodigally than money—indeed, changing the course of civilizational development as happened in the Neolithic and Industrial Revolutions. But while the Neolithic Revolution took more than one millennium, and the Industrial more than a century, Humanity is allowed a mere few decades to accomplish its Ecological Revolution.

"Time is in fact the ultimate limit in the World3 model—and, we believe, in the 'real world'," it says in the book *Limits to Growth: the 30-Year Update*, "Given enough time, we believe humanity possesses nearly limitless problem-solving abilities" (Meadows et al. 2006: p. 223). Experiments with the World3 model show, the longer the world puts off decisive measures to stabilize the environment, the smaller the window for transition to sustainable development. And what could have led to success yesterday, may not deliver results tomorrow.

In 2001 Donella Meadows, once the soul of that small team, passed away. And while she never got to see their final book in print, it was to her, first of all, that the authors owed the humanistic content of the work, including within its context a capacity so seemingly far from their professional sphere of understanding as a discipline capacity for foresight, responsibility, community and love, on which, perhaps, they rested their greatest hope. In the foreword to the final part of this trilogy on the occasion of *The Limits to Growth*'s thirtieth anniversary, Dennis Meadows and Jorgen Randers wrote that they had planned to write one more book, *Limits to Growth: the 40-Year Update*. But plans changed, and there will be no fourth book. As Dennis Meadows acknowledged, there is no point in once again describing a scenario for the future, seeing as by any reasonable allowance, it is a scenario of collapse…

As a final thought, we'd like to say a few words about one other scenario, not studied by the Meadows team but whose possibilities should not be forgotten, particularly in light of the impending ecological catastrophe. According to traditional literary genres, you might call it the classic dystopia. But the likelihood of this dystopian future grows with each passing year. This is the total control of human behavior which, it seems, would allow for the relatively easy resolution of most problems of overpopulation and degradation of the biosphere on Earth.

You could make a person totally controllable, for example, by inserting the corresponding microchip into their body, which in a few decades will be just as simple a procedure as a measles or smallpox vaccination. Such a controlee would acquire offspring for themselves only when considered appropriate by the system controlling them. The system would also relieve them of the charm of hyperconsumption: such desires would simply not arise. At the same time, the controlee would punctually observe all the rules of energy conservation, water economization, etc. and meekly dissolve into nothingness whenever the system required. Unlike technological regulation of the environment, such a program, in the not so distant future, will become not only feasible but not overly extravagant either, and it could be viewed as an entirely acceptable alternative to ecological catastrophe.

With regard to psychological zombification, such kinds of technology (just look at the boob-tube) could today already be considered sufficiently developed and, theoretically, capable of preparing the "human-of-the-masses" to accept this "development path." The problem here is not so much technological or economic, as much as moral and socio-political. Well, and in whose interests will such a mass- "controlling" come into being? It goes without saying: A method will be worked out, and those who wish to possess it for narrow corporate gains will always be found.

By the way, it would be completely delusional to think the described scenario could become humanity's salvation, if, of course, you consider the structure in which this even occurs "humanity." The thing is, not even from a moral standpoint, but from a theoretical one, such complexly organized and tightly synchronized systems do not last long, decaying and collapsing in the briefest periods (there will be more discussion of that in the following chapters). In the case of an artificially created super-totalitarian structure, the time of its existence could hardly go beyond a few decades. So, only a democratic social arrangement with deeply rooted democratic institutions and clearly expressed educational and cultural priorities is capable not merely of defending humanity from this looming threat, but giving us a chance to untangle the web of problems born of the global ecological crisis.

References

Danilov-Danil'yan, V. I. (2003). Sustainable development (a theoretical-methodological analysis). *Ekonomika I Matematicheskie Metody*, *39*(2), 123–135 (in Russian).

Gvishiani, D. M. (2002). The limits to growth—the first report to the clup of Rome. *Elektrony zhurnal Biosphera*, (2). Retrieved from http://www.ihst.ru/~biosphere/Mag_2/gvishiani.htm#_ednref3, (in Russian).

Mayr, E. (1970). *Populations, species, and evolution*. Cambridge: Belknap Press of Harvard University Press.

Meadows, D. (2007, Jan 16). Lecture given at the D.I. Mendeleyev Chemical and Technological University.

Meadows, D. H., Meadows, D. L., et al. (1974). *The limits to growth*. New York: Potomac.

Meadows, D., Randers, J., & Meadows, D. (2006). *The limits of growth: The 30 year update*. London: Earthscan. Retrieved from http://www.peakoilindia.org/wp-content/uploads/2013/10/Limits-to-Growth-updated.pdf

Pestel, E. (1989). *Beyond the limits to growth: a report to the Club of Rome*. New York: Universe Books.

Chapter 8
Programs of Change: Stockholm—Rio De Janeiro—Johannesburg—Rio+20

In 1968, the UN General Assembly adopted the resolution to call a conference on the environment in Stockholm in 1972, which was to discuss a number of problems causing deep distress in world society. Into this discussion went the growing danger of nuclear conflict, the ongoing arms race, and finally the progressive degradation of the environment conditioned on the growth of production and consumption along with rapid increase in the planet's population. Prior to the conference, in 1971, an international seminar on development and the environment at Founex, Switzerland, marked the first way-station on the path to global sustainability. Here it was that experts first announced the existence of an overall ecological threat and of this problem's relevance to third world countries. Most important, the seminar cleared the ground for the Stockholm Conference, which assembled from June 6 to 16 of the following year.

The conference declared publicly what had troubled scientific circles for a long time—that a severe ecological malady had developed not only in isolated regions, but on the planet as a whole. Just as in the first report to the Club of Rome, the proceedings emphasized that civilizational development could not be viewed as separate from the environment and that the two are intrinsically linked. Along with this, it was acknowledged that the course of global development as a way to satisfy humanity's growing needs had entered into deep conflict with the environment, which the computer models in *The Limits to Growth* successfully demonstrated and materials from numerous then unfolding scientific observations, including satellite data, confirmed.

The Stockholm Conference affirmed a Declaration announcing 26 principles by which it recommended the world community be governed. This document gave a first complex look at the issues of peaceful coexistence, economic underdevelopment in the third world, social inequality and the ecological malady. In contrast to ideological and military confrontation, it put forward a thesis on environmental protection for the sake of present and future generations. "In the long and tortuous evolution of the human race on this planet a stage has been reached when, through the rapid acceleration of science and technology, man has acquired the power to

© Springer International Publishing AG, part of Springer Nature 2018
V. I. Danilov-Danil'yan, I. E. Reyf, *The Biosphere and Civilization: In the Throes of a Global Crisis*, https://doi.org/10.1007/978-3-319-67193-2_8

transform his environment in countless ways and on an unprecedented scale. Both aspects of man's environment, the natural and the man-made, are essential to his well-being and to the enjoyment of basic human rights the right to life itself." (Declaration… 1972)

And while the documents and decisions from the Stockholm forum did not have an obligatory character and did not presuppose a procedure for ratification by the various governments, it carried such great resonance that it laid the cornerstone for a wide network of national environmental protection structures and created a powerful impulse for developing environmental legislation in most of the world's countries. Those years were also marked by the establishment of the "green" social movement, which established itself in many governments, one after another. Regarding the direct results of the Conference, it is worth naming the special UN Environmental Programme (UNEP) with its headquarters in Nairobi, Kenya.

In this way, starting from 1972, environmental protection activities have taken on a wide scale, with their primary focus becoming the fight against pollution. Direct expenditures alone on these goals added up to $1.5 trillion over the following 20 years (Danilov-Danil'yan and Losev 2000). Developed countries have invested enormous sums in the modification of so-called "dirty" technologies as well as atomic energy, imagined at the time to be environmentally clean and adequately safe.

However, the disparate, uncoordinated efforts for environmental protection could not dramatically alter the dangerous course of runaway global development. The need for a single program of action for the whole global community was felt ever more acutely. Just such a program would be created, and the formation of the Brundtland Commission, named for its chairwoman, Norwegian politician Gro Harlem Brundtland, served as the first step toward that goal. The commission first gathered in 1983, under the aegis of the UN World Commission on Environment and Development.

The Commission's tasks included preparing proposals for long term strategies in the area of environmental protection, as well as formulating goals that would serve as guideposts as various world governments developed their own frameworks for practical action. In 1987, the Brundtland Commission published a program report under the name *Our Common Future*. A large number of international experts took part in the work, and it was translated into all of the world's most common languages.

Without using the word "crisis," the authors had materially characterized the biosphere as being in a state of crisis, and the planet's demographic situation was described in a similar vein. But, while acknowledging the necessity of specific regulation in the area of natural resource extraction, they put this in relative rather than absolute terms. The measures would depend on the level of technological development and existing social relations. Only under condition of ongoing improvement and control would the opportunity to begin a new era of economic growth open before humanity.

Beyond the dubious nature of this postulate (which we will discuss later), the report did not adequately assess the process of ecosystem disappearance. And the biota was materially equated to an economic resource, albeit one which possesses an ethical, aesthetic and cultural value aside from the monetary.

But if the Brundtland Commission did not proceed to announce the full-blown ecological crisis, another book by leading ecologists and economists did so at the top of its voice: *Environmentally Sustainable Economic Development: Building on Brundtland*, edited by R. Goodland, H. Daly, S. El Serafy and B. von Droste, published by UNESCO, 1991. In it, the contributors said that the global ecosystem served as a sink for the pollution created by the economic subsystem. However, under the weight of the latter's extreme growth and expanding size, this pollution has become too great relative to the biosphere. As a result, the absorbency of biospheric sources and sinks has come under unrelenting stress. And while, in the recent past, a person could go about their business without a thought to the adaptive capabilities of the biosphere, and the world seemed a bottomless reservoir, able to swallow up any amount of economic byproducts, now the era of the "empty world" has come to an end, and the "full world" epoch has begun (Environmentally...1991).

***

Both the Brundtland Commission Report and *Building on Brundtland* were lying on the table among the working materials at the UN Conference on Environment and Development (UNCED), held from June 3 to 14, 1992. It was a truly global conference, unlike anything history had seen. Representatives from 174 countries took part in the Rio Summit, including 114 heads of government or state, 1600 nongovernmental organizations and a countless number of journalists. At the same time in Rio, the "Global Forum" on environmental problems was going on, drawing about 9000 different organizations and 29,000 individual participants as well as 450,000 guests and observers arriving on their own initiative. Thus, this event fully earned the right to be called the Earth Summit (United Nations Conference... 1992).

One unquestionable achievement of the Conference was its accompanying intellectual process, broad discussion and exchange of ideas, over the course of which the whole body acknowledged the strategically significant postulate that problems of the environment and development could not further be viewed separately. The UNCED convincingly demonstrated the organic interrelation between the state of the environment, poverty and underdevelopment among a significant portion of third-world countries, and the vicious system of production and consumption in most developed states. The pressure of population growth on nature, energy use and climate change, the tropical lumber trade and desertification—all these aspects of global and regional ecology were discussed at such a level and attracted attention of such scales as would hardly have been dreamt of at the time of Stockholm.

But, we dare say, the most meaningful result of the Conference was the widespread introduction of the term Sustainable Development, which was conceived as an alternative to the previous, nature-destroying course of civilization. Here is how the Brundtland Commission formulates and interprets the concept: "Sustainable development is development that meets the needs of the present without compromising the ability of future generations to meet their own needs" (Our Common Future 1987: p. 41).

The UNCED's proposed concept of sustainable development was based on the Brundtland Commission report and included the following main proposals:

- The main priority of sustainable development should be people, who have a right to a healthy and productive life in harmony with nature;
- Environmental protection should become an inexorable component of the development process and cannot be viewed separately from it;
- The task of preserving the environment involves not only the present generation, but future ones as well;
- Reducing the gap in standard of living between countries and eliminating poverty and want are among the most important tasks of the global community;
- In order to transition onto the path of sustainable development, governments should re-examine models of production and consumption that do not facilitate it (Rio Declaration… 1992).

Over the course of the Conference, the assembly adopted several documents, the most important of which were the Rio Declaration on Environment and Development and Agenda 21 (A plan of action for achieving ecologically sustainable development going into the twenty-first century).

The Declaration reflected the evolution of thought concerning environmental problems over the 20 years since the Stockholm Forum. These ideas, or principles, were recommended as guidelines to develop plans for a transition to sustainable development, addressed to the whole global community as well as the various states.

So, for example, Principle 1 postulates the leading role of the population in realizing sustainable development. The state serves as guarantor for environmental quality and carries responsibility to other countries for any harm done (Principle 2). Principles 3–5 particularly emphasize the inseparability of socio-economic development goals from the interests of environmental preservation for both present and future generations. Principle 10 asserts the major significance of the public's role in resolving environmental problems, while Principle 11 does the same concerning the development of environmental legislation.

A special article, Principle 15, focuses on ecological caution: "Where there are threats of serious or irreversible damage, lack of full scientific certainty shall not be used as a reason for postponing cost-effective measures to prevent environmental degradation." Furthermore, governments are advised to use economic mechanisms as a means of protecting nature, including payments for pollution (Principle 16), as well as an ecological expertise mechanism to assess harmful environmental consequences of planned activities (Principle 17), notifying other states of natural disasters and technological accidents fraught with cross-border consequences (Principle 18) and so on (Rio Declaration…, 1992).

The other most important document of the Summit was Agenda 21. While in the Stockholm Plan of Action, the great majority of recommendations related to five problems (environmental assessment and management, detecting global pollution, environmental education, information and culture, development and the environment), with Agenda 21, the accent was put on social and economic development, justice and international cooperation. The "Agenda" includes over 100 programs covering a wide range of problems, from overcoming poverty to strengthening the role of the public in resolving ecological challenges.

A few important aspects, however, such as the structure of consumption, debt in developing countries or the export of dangerous waste, were either poorly represented or absent entirely from the authors' purview. Nonetheless, Agenda 21 served as a kind of touchstone for national programs of transition to sustainable development, which UNCED encouraged all the world's governments to develop. At present, no fewer than a hundred countries have them. (In Russia, such a program has still not been adopted, though a project for one was developed way back in 1997.)

We cannot avoid speaking, however, of the other side of the coin, the sense of disappointment with the results of the Rio Summit, which was unable to rise to the level of the challenges standing before it. Of particular notice among the general choir of criticism rang out the voices of such authoritative specialists as Ernst von Weizsacker, Herman Daly, Donella and Dennis Meadows and many others.

More than anything else, it was several of the outcome documents from that landmark forum that left them deeply dissatisfied. Despite documenting global changes to the environment—the elimination of forests, the reduction of the biosphere, the dangerous climate shifts—none of them acknowledged the fact that the planet had truly entered a phase of full-scale ecological crisis, and that this crisis demanded a radical re-thinking of the existing principles of global development. And, most importantly, no attempt was made to initiate development of a scientifically-based strategy for such development and lay for it a solid theoretical foundation. Just the opposite, more likely, in their understanding of the problems at hand, the majority of conference participants came from a position of pure criticism, centered on the store of past experience. And this experience, it seemed, had demonstrated more than once the broad human capacity to untangle the tightest imaginable knots with help from the achievements of scientific progress or improvements to social and economic institutions[1]. By counting on such, as it would seem, tried-and-true instruments, the conference participants were also clearly trying to apply that proven previous experience to the present day. They guessed upon an answer to a question fundamentally new to civilization with the help of structural and technological reconstruction of industry, introduction of low-waste technologies and other well-worn approaches from past decades (Danilov-Danil'yan and Losev 2000).

And, meanwhile, in the second half of the twentieth century, humanity made a discovery of such magnitude that, against its backdrop, thinking in the same categories has become impossible: *Humanity has "discovered" the environment for itself.* After centuries of ignoring it as something external, having only an indirect relationship to himself, man suddenly found that the environment is connected to every

---

[1] We recall in this regard the Franklin Roosevelt presidential administration's series of reforms in the areas of labor protections, social services, tax collection and banking, which radically changed the socio-economic landscape of the United States, lifting the country in a few years from the grips of a most severe crisis. Or post-war West Germany, magically reborn from the ruins thanks to American financial aid and a thoroughly thought-out economic and social program led by economic minister Ludwig Erhard (the "German economic miracle" as it is called).

aspect of his being without exception and in the most intimate ways, from the global economy to the state of his health. And that the functioning of the environment obeys its own intrinsic laws which people have lagged behind a hundred years in studying, entering, as a result, into an intractable conflict with nature. It was this inherently new reality, clearly, that a majority of conference participants failed to take into account, mechanically superimposing all that had been worked out in the preceding century and a half onto today's fundamentally different situation.

***

In accordance with decisions in Rio de Janeiro, the following World Summit on Sustainable Development was to take place in 10 years. After about 3 years of planning, it opened in August, 2002, in Johannesburg, South Africa, and nearly matched the Rio Summit in representativeness and number of participants. But while UNCED-1 stirred hope among the global public, the same, unfortunately, could not be said of "Rio+10," in large part due to the extremely limited progress made in the area of sustainable development over the previous decade. This skepticism proved justified, and grounds for disappointment were more than adequate. As the Johannesburg Declaration on Sustainable Development said on this account: "The global environment continues to suffer. Loss of biodiversity continues, fish stocks continue to be depleted, desertification claims more and more fertile land, the adverse effects of climate change are already evident, natural disasters are more frequent and more devastating, and developing countries more vulnerable, and air, water and marine pollution continue to rob millions of a decent life" (Johannesburg Declaration… 2002).

Two main documents were adopted at the Summit—the previously mentioned Declaration and the Johannesburg Plan of Implementation, a high-level action plan concerning sustainable development.

In both documents, a leading position was devoted to poverty as the main factor of social instability, crime and moral decay. Poverty gives rise to a sense of hopelessness and apathy, and with it, total irresponsibility in relation to nature, society, and, finally, to one's own children who, finding themselves at the social bottom, receive a perverse conception of the surrounding world, in turn becoming antisocial personalities. At the same time, poverty is inseparably connected to weakness in the economy and the issue of work. Unemployment, after all, is one of the main factors in social degradation. But as most job openings in modern production, and even more so in management, require education and qualifications, access to them is an irreplaceable precondition to eliminating poverty. In Johannesburg, therefore, it was proposed that states develop national programs to provide wider access for poor citizens to productive resources, credit and education, as well as equality for all members of society in receiving education or work (Marfenin 2006: pp. 583–84).

The plan contained a number of other important recommendations from the social and environmental spheres: providing the poorer classes with access to agricultural resources, including free introduction to sustainable farming methods; transfer of affordable energy technology to developing countries (biomass, wind generators, small hydroelectric stations, etc.); development of rational economic methods to prevent degradation of land and water resources; and so on

(UN Johannesburg Plan... 2002). The Summit also recognized the need for dramatic changes in the established system of production and consumption and called on countries to encourage models that would not bring harm to the environment or undermine the natural resource base. For the first time at this level, the problem of globalization was looked at with all of its positive and negative consequences for various countries and world regions.

But while Rio and preparatory work leading up to it made first, uncertain steps toward the creation of a scientifically-based concept of sustainable development, the Johannesburg Summit preferred to avoid looking at questions of that sort. Concentrating on isolated, albeit extremely relevant problems of modernity; such as the fresh water deficit, food supplies, energy and preserving biodiversity, it was as if it had demonstrated with its whole attitude that resolution of humanity's pressing issues required not so much plans and programs as incessant undertaking of concrete practical steps. World Summit General Secretary Nitin Desai acknowledged in his speech that the participants did not foresee any great breakthroughs or the signing of any treaties (Johannesburg High Level...2002).

Indeed, many of the agreed upon target indicators had been confirmed at lower level functions—during development of the Millennium Development Goals, adopted in accordance with the UN General Assembly decision of September 8, 2000, and in the execution of its Millennium Declaration (UN Millennium Declaration... 2009). The main attention of the Summit was focused on working out diverse concrete plans, goals and graphs. Desai, in his closing words, recognized that many attendees would have wanted more meaningful results, but that achieving them would require additional resources (Johannesburg High Level... 2002).

But people were waiting for a breakthrough from the Johannesburg forum, or at least a serious strategic layout for the future development path insofar as its very Declaration acknowledged that the world was not approaching sustainability, but rather was moving further from it. Were the issues placed on the agenda important? Yes, unquestionably important; on their successful resolution depends the well-being of tens of millions of people. But trying to solve each separately, without regard for their systemic interaction, is obviously a futile business. After all, a conference at such a level doesn't assemble every year. It's an event of global significance. You could say without exaggeration that the world awaited some fateful decision, where the most relevant question standing before humanity is "to be or not to be?" and the majority of global environmental indicators demonstrate a sustained trend of decline. But, unfortunately, the summit could not rise to the level of its own mission.

***

If the conference at Rio de Janeiro (1992) and the Summit at Johannesburg (2002) met with no few public expectations, then the Rio+20 Summit, assembled in the Brazilian city for the twentieth anniversary of the Conference on Environment and Development, remained under a shadow and certainly did not become an event. This, despite the fact that its declaration "The Future We Want" and other outcome documents contained a particular novelty.

In part, this included the idea of a "win-win," simultaneously addressing socio-economic and environmental problems. As the experience of past decades had shown, attempts to resolve them separately fail to engender an interested response from civil society and, consequently, do not lead to success. Therefore, programs directed at raising employment or improving people's quality of life were recommended to immediately include corresponding environmental priorities. In other words, socio-economic projects should involve the resolution of ecological problems, and environmental projects—provide a positive socio-economic effect. In this way, people's interest in the resolution of issues troubling them would draw them to address environmental problems as well.

Among the concrete proposals and plans adopted by Rio+20, we will note the UN General Secretary's stated aim of developing the concept of energy security, raising efficiency in forest management and the creation of new development goals meant to replace the old Millennium Development goals in 2015. Beyond this, a number of important agreements were made in back rooms of the forum relating to the financing of sustainable development projects in the areas of agriculture, energy, transport and forest restoration.

And nonetheless many participants of the forum expressed their dissatisfaction with its results. It had done much less than it could have, and the basically correct declarations turned out untethered to concrete practical steps and corresponding legal obligations, not to speak of the fact that no agreement was reached to adopt obligations for ocean resource protection or any progress made on the issue of removing fossil fuel subsidies (Pisano et al. 2012).

As a result, the social organizations represented at Rio de Janeiro, having come forward with the petition under the name "The Future We Want," disassociated themselves unanimously from the outcome documents of Rio+20. They cast particular attention on the lack of progress in water resource management which, in the opinion of WWF director Lasse Gustavsson, should be based on natural rather than political limitations. "What we need is…a duty to protect and restore natural drinking water supply systems, forests, which protect water resources, and to prepare the world for the hits it will take from climate change" (RIA.ru 2012). Kevin Henry, project coordinator for "Where the Rain Falls," published an article called "Rio Plus 20 or Rio Minus 20," judging the 2012 conference as a giant missed opportunity, or even a large step backward (Care International 2012).

***

Despite all of the grievances lodged at UNCED 1992 and the following global forums on the environment and development, each of them has become a milestone on humanity's historical journey. In the direction of increasing global sustainability, we might like to add. But, unfortunately, objective data testifies to something else. The world is still moving in the direction of unsustainability. That is what sets the stakes so high. Indecision and half measures on environmental protection, after all, do not amount to running in place, but rather to an inevitable slide toward global catastrophe.

As Kevin Henry said in his above-mentioned article, "Rio Plus 20 or Rio Minus 20," "The clock really is ticking, not least because of the threat posed by climate change, but our political leaders—almost universally—do not seem to hear it. Or worse yet, the political elite hear it and choose to ignore it, thinking that making major changes in our approach to development can wait until they have attended to other 'more important' or 'more pressing crises." (Care International 2012). Recalling that since 1992, when the first Rio Summit was held, global CO_2 emissions have grown 40% and biodiversity has fallen by 10%, he considered it necessary to add that at current pollution levels global warming, as scientific data testifies, "will continue unabated and almost certainly exceed the 2 degrees centigrade deemed 'safe'."

Thus, it is from these positions that we must approach assessment of all four international forums. In truth, it will be nature itself that judges them most harshly, having, despite many unquestionable local successes in its protection, demonstrated a sustained tendency toward degradation across a whole range of global parameters. Against this backdrop, the only major achievement the world community can truly be proud of is the ozone story, which we discussed in the previous chapter. Even there, it is still premature to speak of ultimate stabilization of the ozone layer, despite a full cessation in production of ozone-destroying substances.

Granted, we cannot underestimate the significance of this victory: After all, beyond the direct, physical result, humanity received convincing confirmation that global ecological projects could be realized. However, this lone success has not yet been followed up by any other such tangible headway, and there is no question that the world is becoming ever more ecologically unsustainable.

Unfortunately, this last fact also correlates to a noticeable decline in public enthusiasm for the idea of sustainable development, whose peak arrived in the 1990s. Furthermore, after the terrorist attacks of September 11, 2001, in New York, after the wars in Afghanistan and Iraq, after the financial crisis of 2008–2009, after the series of social earthquakes in the Arab World and the wave of immigrants pouring into Europe, one also notices a decline in the number of scientific publications focused on the state of the environment. That's understandable: The world community simply doesn't have the strength or the money to deal with everything. Thus we are forced to make an unconscious choice between the pressing issues of today and those that may come to fruition the day after tomorrow should we put off the resolution of long-term problems for later.

This is a false dichotomy, however, like choosing between the health and wealth of one's children and grandchildren. Sustainable development represents not only a life of peace and harmony with nature. It also entails the population health of humanity, its social and interethnic stability, and the rebirth of many age-old values common to the human race, lost or deformed through the costs and distortions of modern civilization. Thus, in putting off the resolution of long-term problems, we may arrive at "tomorrow" with empty hands, when it is too late to solve them.

References

Danilov-Danil'yan, V. I., & Losev, K. S. (2000). *The ecological challenge and sustainable development*. Moscow: Progress-traditsia. (in Russian).

Declaration of the United Nations Conference on the Human Environment. (1972). *From report of the United Nations conference on the Human Environment*. Stockholm. Retrieved from http://www.un-documents.net/unchedec.htm

Goodland, R. J. A. (1991). *Environmentally sustainable economic development: Building on Brundtland*. Paris: UNESCO.

Henry, K. (2012). *RIO 20 Plus 20 or minus 20. CARE International*. Retrieved from https://www.care-international.org/news/stories-blogs/rio-20-plus-20-or-minus-20

Johannesburg Declaration on Sustainable Development World Summit on Sustainable Development. (2002). Retrieved from http://www.un-documents.net/jburgdec.htm

Johannesburg High Level Meeting. (2002). Retrieved from https://www.un-ngls.org/orf/pdf/ru96.pdf

Johannesburg Plan of Implementation of the World Summit on Sustainable Development. (2002). Retrieved from http://www.un.org/esa/sustdev/documents/WSSD_POI_PD/English/WSSD_PlanImpl.pdf

Marfenin, N. N. (2006). *The sustainable development of humanity*. Moscow: MGU. (in Russian).

Our Common Future: Report of the World Commission on Environment and Development. 1987. Retrieved from http://www.un-documents.net/our-common-future.pdf

Pisano, U., Endl, A., & Berger, G. (2012). *The Rio+20 conference 2012: Objectives, processes and outcomes. ESDN Quarterly Report 25—June 2012*. Retrieved from http://www.sd-network.eu/quarterly%20reports/report%20files/pdf/2012-June-The_Rio+20_Conference_2012.pdf

RIA.ru. (2012). WWF has criticized the draft Declaration of the upcoming UN Forum "Rio+20". Retrieved from https://ria.ru/eco/20120112/538243479.html (in Russian).

Rio Declaration on Environment and Development: Report of the United Nations Conference on Environment and Development*. (1992, Rio de Janeiro, June 3–14). Retrieved from http://www.un.org/documents/ga/conf151/aconf15126-1annex1.htm

U.N. General Assembly. (2009). *United Nations Millennium Declaration Resolution adopted by the General Assembly 8 September 2000*. Washington: World Bank.

United Nations Conference on Environment and Development. (1992). Retrieved from http://www.un.org/geninfo/bp/enviro.html

Chapter 9
The Path to a Systemic Understanding of the Biosphere

Like a little worm chewing its way through its chosen apple, man has built his civilization within the biosphere at the cost of its partial destruction. But while the larval codling moth, reaching maturity, deserts its devoured fruit, humans lack the opportunity to do the same and, abandoning our "apple," settle other planets. It was not long ago at all that we as people began studying this most complicated of systems. The first attempts at a universal, holistic approach to the biosphere—long before the term itself appeared—arose when Alexander von Humboldt began his work. It was Humboldt (1769–1859) who counterposed the mosaic of independent organisms proposed by Karl Linnaeus with the concept of an interrelation of organisms between each other and the landscape, laying the basis of biogeography. Nonetheless, by the second half of the nineteenth century, Humboldt's views of a united earth system with a strong influence of climate upon the living world had made way for the historical descent of organisms (Phylogeny) as the lone scientific explanation for natural phenomena deserving of attention (Zavarzin 2004).

Alexander Humboldt

Charles Darwin used the history of descent, through a process of competitive natural selection on the basis of variability and persistence of successful mutations among offspring in response to tasks of adaptation to environmental conditions, to explain the linear diversification of species. Darwin's theory, convincing in its logic and freed of the necessity of appealing to external forces to explain biological diversity and the sustainability of species, became, however, less a theory of evolution and more a paradigm shift in world views. Within the bounds of its subsequent development, a reductionist approach came to prevail in biology—an explanation of the whole by way of the parts on the basis of acquired empirical material—which focused scientists' attention on the evolutionary fate of isolated species and individual specimens, gradually decompartmentalizing the biome. This tendency, taken to its extreme, seriously delayed the development of views on the biosphere as a unified system with all the rules of a whole. As a result, by the turn of the twentieth century, only a few minds hazarded to approach research of the biosphere from this point of view.

You might think that a systemic concept of the biosphere should have arisen as part of the then-emerging field of ecology, but, as it happened, everything happened differently. And the first to arrive at the modern understanding, by his own, independent path, was not a biologist but a mineralogist: the founder of geochemistry, prolific Russian scientist Vladimir Vernadsky (1863–1945). He, in turn, based his work upon that of his great predecessor, founder of soil sciences Vasily Dokuchaev (1846–1903). In a set of lectures published by Vernadsky under the title "*Biosfera*" (The Biosphere) and released 3 years later in French (*La Biosphere* 1929), he put forward the idea of a holistic world in which living material ("the membrane of life") is connected through a system of biogeochemical cycles in the atmosphere, hydrosphere and lithosphere. He proposed that we call this covering of the Earth, in which all biogeochemical processes run their course, the biosphere.

Vladimir Ivanovich Vernadsky

Vernadsky showed that the chemical state of the Earth's crust lies entirely under the influence of life and is determined by living organisms. His studies not only looked at the basic qualities of life materials and influences on them by chemical compounds, but also first explored the reverse influence of life upon the abiotic medium with the formation of such bio-inert natural bodies as, for example, soil. For the first time, Earth's covering was conceived as a single, complicated, and at once fragile entity. As he put it, the process of its evolution is expressed in the natural bio-inert bodies that play a foundational role in the biosphere—soils, surface and ground waters, anthracite and bitumen, limestone, nutrient minerals, etc. (Vernadsky 1998). In the monograph "The Chemical Construction of the Earth's Biosphere and its Surroundings," also published posthumously, he directly calls the biota an enormous geological force: "*Living organisms* are *functions of the biosphere* and connected to it tightly in both matter and energy, and are an enormous *geological force which determines it*" (Vernadsky 1987, p. 45).

Along with this, thinking on the paths of evolution of the biosphere and the special place that humans occupy within it, Vernadsky came to the idea of possibly governing the biosphere through the power of human reason. "We are presently living through an exceptional phenomenon of life in the biosphere, connected genetically to the appearance, hundreds of thousands of years ago, of *Homo Sapiens*, by this path creating a new geographical force, scientific thought, sharply increasing the influence of life material in the evolution of the biosphere. Completely overtaken by life material, the biosphere increases, clearly, its geological force to unlimited size and, processed by the scientific thought of Homo sapiens, transitions to a new state—the noosphere" (Vernadsky 1988, p. 32).

In this sense, Vernadsky was a man of his time and age, bound by hope in the future and the limitless, as it then seemed, possibilities of scientific progress. But we've already come to a different aspect of Vernadsky's legacy—his widely-known idea of the noosphere (the sphere of reason, the human "thinking membrane" of the planet), which we will settle on in more detail in Chap. 16.

***

Vernadsky's ideas, coming far ahead of their time, could have long remained abandoned if not for a new field that was speedily developing at roughly the same time—ecology—which focused the attention of scientists on the structure and particular

functions not of isolated organisms, but of the biological complexes they make up. Though ecology owed its establishment mainly to existing biologists, the two fields did not truly come to agreement until the second half of the twentieth century. And while a first understanding of ecology was proposed by the famous German naturalist, philosopher and Darwin-supporter Ernst Haeckel (1834–1919) to distinguish the area of biology which studies interaction of organisms with the environment (he called it "the economics of nature"), the term was hardly ever used in scientific circles until the early 1900s. Hydro-biologists made a particularly significant contribution to the establishment of this new branch of science, which is understandable: water ecosystems (especially reservoir ecosystems), as a rule, are easier to wall off. By their very nature, it seems, they are isolated from surrounding ecosystems.

Ernst von Haeckel

Among the first specialists in ecology stands German zoologist Karl Mobius (1825–1908). While studying mollusk reproduction in North Sea oyster beds, he confirmed the existence of an internally linked community of organisms inhabiting one or another identical portion of sea floor, which he called a biocenose (1877). At the same time, Mobius noted definite adaptations acquired through evolution, the attachment of given species not only to each other but also to specific conditions of the local abiotic environment—the *biotope*. As a result, the concept of biocenose was applied to freshwater communities as well—the biocenose in a pond or lake. Then, also to land—the biocenose of a birch forest, a riverine meadow, etc.

Karl August Mobius

But truly widespread study at the supra-organism level began in the early twentieth century, with biologists from many different backgrounds—botanists, zoologists, hydro-biologists, forestry specialists, etc.—each making their contribution. They considered it particularly important to discover a set of general rules, which would characterize the development of the most diverse organism complexes (communities, biocenoses) in the course of their interaction with the environment. That would include, for example, the process of *ecological succession*, the regular stage of development for the most diverse type of ecosystems.

The discovery of succession was the work of two American botanists. The first, Henry Cowles (1869–1939), conducted research of vegetation on the shores of Lake Michigan, which over a long historical period had slowed and retreated from the shoreline. He correctly hypothesized that the growth of a community should increase in proportion to its distance from the tide, and, in this way, was able to reconstruct a detailed scheme for the whole process. The youngest, just-formed sand dunes were seeded with perennial grasses that put down roots in shifting sands. Then taller grasses would appear in their place, followed by shrubs. Under this formed canopy, on the older and more established dunes, trees would start to grow, in a strictly determined order of succession: first pines, and after a generation, oaks and maples would replace them. Finally, furthest from the shore, there appear beech trees—the most shade-loving trees for that climate (Odum 1983).

Illustration: Journal "Nauka i zhizn" 2010, No. 3

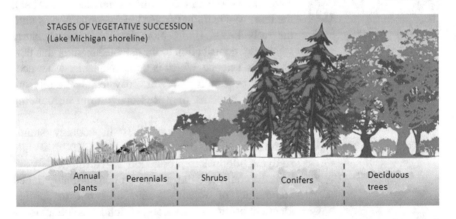

In 1916, Cowles adherent Frederic Clements (1874–1945) published his classic work *Plant Succession*. Viewing the vegetation community as a single, holistic organism undergoing degrees of development from infancy to maturity, he showed the adaptability of biocenoses, their ability to adjust and evolve as the environment changes. While at the early steps various communities on the very same place may differ greatly from one another, at later stages they become more and more similar. It turns out in the end that for every area with a particular climate and soil, there is only one characteristic, mature, in Clements' terms *climax community*.

Ten years later, in England, zoologist Charles Elton (1900–1991) released the book *Animal Ecology* (1927), which established the field of population ecology and allowed zoologists to switch their attention from the isolated organism to the population as a whole, the independent unit level at which specific particularities of ecological adaptation and regulation appear. The author, who had recently been on two Arctic expeditions, took an interest in cyclical variations in the number of small rodents that occurred every 3–4 years. Having observed many years' worth of data from the North American fur trade, he came to the conclusion that hare and lynx also demonstrate cyclical variation, though their numbers peak roughly once in 10 years. In this work, also considered a classic, the structure and distribution of animal communities are first described, and, furthermore, Elton introduces the concept of an ecological niche and formulates the rules of an *ecological pyramid*—the consecutive lessening of the number of organisms from the lowest trophic levels to the highest (from plants to herbivores, from herbivores to predators and so on) (Elton 1946).

The 1920s and 30s were marked by the introduction of precise research methods into ecology, led by mathematicians American Alfred James Lotka (1880–

Charles Elton

Alfred Lotka

1949) and Italian Vito Volterra (1860–1940). In Lotka's book, *Elements of Physical Biology*, released in 1925, the first attempt was made to use quantitative methods in the field of biology. In part, Lotka developed mathematical models for interaction between species (for example, a model showing the inter-connected trends in the numbers of predators and prey) as well as biogeochemical cycles. While Lotka never used the term "ecology," his attempts to apply the laws of physics to biological study clearly illustrates the tendency to expand the field of research conducted as part of ecology (Lotka 1925). In 1926, Volterra developed a mathematical model for competition between two species for one food source and showed the impossibility of their extended sustainable coexistence.

The theoretical research of Lokta and Volterra attracted the attention of young Soviet biologist Georgy Gause (1910–1986), who presented his own modification of the equation, more cogent to biologists, describing the processes of interspecies competition. His experimental tests of the models, conducted with laboratory cultures of bacteria and protozoans, showed that species coexistence is possible only if it is determined by distinguishing features of the environment, i.e., when the species occupy different ecological niches. Among competitors for the same niche, species inevitably push each other out (the competitive exclusion principle). Gause's work was published in the US in

1934 as *The Struggle for Existence*. It only saw the light of day in Russia seven decades later. In many ways, the book facilitated the emergence of population biology. The emphasis it placed on trophic connections as the basic path for the flow of energy through natural communities made a major contribution to the nascent concept of ecosystems.

The honor of introducing this concept, however, belongs by rights to English *botanist* Arthur Tansley (1871–1955). Of course, he had his own highly authorita-

tive predecessors, of which we might name American hydro-biologist Edward Birge, who researched the role of lake organism communities in mineral cycles and transformation of energy, or his German colleague August Thienemann, who in the 1920s formulated such important concepts for ecology as biomass and biological production. But, nonetheless, it is 1935 that ecologists consider the year of birth for their field as an independent branch of science. Tansley's main achievement was to successfully integrate the biocenose and biotope into a new function unit—the *ecosystem*. And while other, more established sciences, such as physics, chemistry or cell biology had long possessed their own basic unit—atom, molecule, cell—now ecology had the ecosystem: A single natural complex limited in

Arthur Tansley

time and space, created by living organisms and their environment, where living and inert components are linked by mineral exchange and the distribution of energy flows.

In 1942, independently of Tansley, Russian biologist Vladimir Sukachyov (1880–1967) developed the concept of biogeocenose based on forest communities. Generally

analogous to an ecosystem (synonyms, really, and many ecologists use the similar term landscape), the biogeocenose is characterized by limited geographic extent and homogeneous natural and climactic conditions. On land this could be a small plot—a subsystem of the landscape (such as a riverine meadow with the soil beneath it and canopy above), including the biotic and abiotic components of the environment united by a mineral cycle and flow of energy. Both territorially and hierarchically biogeocenoses can be viewed as the units or "cells" of the biosphere, which, in turn, is itself an ecosystem of a higher level—the global ecosystem of Earth (Reymers 1990).

The appearance of the ecosystem as a concept sharply changed the situation in ecology, which had noticeably suffered from overextension over the vari-

Vladimir Sukachyov

ous branches of science, and laid the groundwork for a wide arena of ecosystem research. As before, here hydro-biologists played a leading role. Their specialization—aquatic organisms often dwelling in closed reservoirs (ponds and lakes)—

being distinguished by the tightly weaved interconnection of physical, chemical and biological processes.

So the above-mentioned limnologist Edward Birge, studying the "breathing of lakes," through strict quantitative methods, was able to establish the seasonal trends of dissolved oxygen content, which depends not only on the agitation of water mass and oxygen diffusion from the air, but also on the activities of organisms that produce oxygen (plankton, algae) or use it (bacteria and animals). As a result, these ideas were developed in the works of Russian limnologists Leonid Rossolimo (1894–1977), Georgy Vinberg (1905–1987) and others. Vinberg developed the energy balance approach, allowing further research into the mineral cycle and transformation of energy in an ecosystem on the basis of purely quantitative indicators. According to his method, one used the unity of biochemical processes taking place in the various organisms—such as photosynthesis in algae or all plants in a forest— to add up the results of their activity according to the quantity of organic material and free oxygen formed thereby. In this way, the opportunity arose not only to place a quantitative value on biological production by forest or water ecosystems, but also to design theoretical mathematical models based on the energy approach.

Three years later in the US, George Hutchinson (1903–1991) established similar

Raymond Lindeman

methods, collecting his own research and that of other scientists into his *Treatise on Limnology* (1957), which still represents the most complete summary of lake-borne life in the world. For this reason, his school of thought greatly influenced the development of ecology in many countries. First among his students worth noting is Raymond Lindeman (1915–1942), who sadly passed well before his time. His short work, *The Trophic-Dynamic Aspect of Ecology,* (Lindeman 1942), without exaggeration, brought about a new era in ecology. Scientists from all over the world still cite it to this day. In this work, Lindeman developed a general scheme for the transformation of energy in an ecosystem and laid out the basic methods for calculating the balance of energy. In part, he theoretically demonstrated that during the transfer of energy from one trophic level to another (from plants to herbivores or carnivores), the quantity of energy is reduced. Thus an organism of each consecutive level has access to only a small part of the energy, no more than 10%, which belonged to the organisms of the previous level.

Since that time, ecosystem research has become one of the main currents in ecology, and the quantitative determination of components in ecosystems—one of the principle methods that allow us to model biological processes.

***

Thus, step by step, by the efforts of hundreds of scientists, ecology pieced together the incomplete fragments of the construction and occupied the structure whose vaults and contours Vernadsky had described in his works. However, the field had not yet risen to an understanding of the biosphere as a global system.

Vernadsky died in the final year of the Second World War, and his ideas remained undervalued in many ways by his contemporaries. Even his magnum opus, a type of scientific inheritance, "Chemical Composition of the Earth and Its Environs," was only published 15 years after his death. It took still another decade for scientists to confirm his view of the biosphere as a single holistic system. General systems theory, associated with Austrian biologist Karl Ludwig von Bertalanffy(1901–1972) in the 1940s, played a role in this. Bertalanffy studied mathematical rules for different types of systems under the most general view. It was Bertalanffy who introduced the concept of an open system (as opposed to a closed one, whose many diverse variations are studied in theoretical physics), which distinguished the specifics of living organisms existing on a constant flow of matter from the environment. These provide themselves with additional energy, enabling a lowered level of entropy and creating the preconditions for sustainability of living systems in relation to the environment.

Among the number of Russian scientists who followed Vernadsky's line, it's worth mentioning first and foremost, the remarkable biologist Nikolay Timofeyev-Resovsky (1900–1981). Having made his mark during the interwar decades, when he conducted research into radiation genetics in Germany, in his later years Timofeyev-Resovsky focused on issues of global ecology. In many ways, he anticipated current understandings of a wide number of environmental problems which were then only just emerging. In the report, "Biosphere and Humanity," that he made in 1968 at a division meeting of the Obninsk City Geographical Society, where he lived after release from the GULAG (Moscow, Leningrad and other large cities being closed to him), he compared the biosphere to a giant living factory, reshaping matter and energy on our planet's surface.

Nikolay Timofeyev-Resovsky

The biosphere, according to the report, "forms the balanced makeup of our atmosphere, the diluted makeup of natural waters, and, through the atmosphere, the energy of our planet. It influences the climate. Recall the enormous role of water evaporation for vegetation and the moisture cycle on the Earth, the vegetative cover of Earth. Therefore, the Earth's biosphere forms all of man's surroundings…To sum up, without a biosphere or with a poorly working biosphere, people cannot exist on Earth" (Timofeyev-Resovsky 1996, pp. 59–60).

This report, in the form of an article by the same name, was printed in a collection of scientific works by the Obninsk Department of the Geographical Society. But given the specifics of this obscure publication, few read it. Fewer still, perhaps enough to count on one's hands, could see the value of the scientist's innovative ideas. As so often happens with Russian trailblazers, both report and article passed by nearly unnoticed. Nor did the Academy of Sciences at that time care to remark on the fallen scholar. But here Timofeyev-Resovsky had almost first expressed a very important idea about the environment's full-scale management of life on Earth.

Unfortunately, being on this side of the "iron curtain" often put Russian scientists in a notedly disadvantageous position, and the ideas that Timofeyev-Resovsky expressed remained truly beyond the field of vision for Western scholarship.[1] Instead, an unusual degree of interest in the scientific world was aroused by a different biospheric conception, put forward in the 1970s by English scientist James Lovelock (1919–). He called it "Gaia," after the Greek goddess of Earth.

James Lovelock

An engineer by education, Lovelock had previously worked at NASA, where he designed tools for the discovery of life on other planets in connection with future flights by automated stations to Mars and Venus. Even earlier, as a university student, he created a unique gas spectrophotometer for the measurement of minute concentrations of gases in the atmosphere. It was using precisely this tool that scientists managed to detect increasing quantities of chlorofluorocarbons destroying the Earth's ozone layer. This professional activity led Lovelock to the idea that the existence of life on a planet could theoretically be detected according to the makeup of its atmosphere as the most volatile environmental medium, the most sensitive to any biogeochemical changes. The atmosphere of a "living" planet, Lovelock proposed, should be distinguished by a thermodynamic disequilibrium supported by life activity. By the same token, a "non-living" planet has an atmosphere whose makeup is determined by the average chemical composition in a state of equilibrium. All of these considerations spurred the further formation of his hypothesis, best known as the Gaia Principle, which was first published in the form of an article, then developed into a number of books and monographs.

The image of Gaia, according to Lovelock, arises as one looks thoughtfully upon our planet from space, when it is seen as a complex, multi-level living organization. Or when mentally travelling from the macro-level to the micro: biosphere> biocenose> organism> organ> cell. The whole shape of the Earth, he writes, "The climate, the composition of the rocks, the air and the oceans, are not just given by geology; they are also the consequences of the presence of life. Through the ceaseless activity of living organisms, conditions on the planet have been kept favourable for life for the past 3.8 billion years. Any species that adversely affects the environ-

---

[1] Which, by the way, one might attribute to his not winning the Nobel Prize. He entirely could have shared the prize won by his younger colleague Max Delbruk, with whom, at one time in early 1930s Germany, he carried out the work of determining the size of a gene.

ment, making it less favourable for its progeny, will ultimately be cast out, just as will those members of a species who will fail to pass the fitness test" (Lovelock 1991, p. 25). Gaia is imagined as some kind of self-organizing system, like a "super-organism" possessed of self-regulating "geophysiological" properties and maintaining the global environmental parameters through homeostasis at levels favorable to life. Evolution of the biota is so closely linked to that of its physical environment that together they form a single self-perpetuating system, by its nature recalling in part the physiology of a living organism.

In his configuration, Lovelock gives particular attention to the Earth's bacterial community, whose role in the evolution of the biosphere from the first appearance of life to our time hardly requires proof. Bacteria, after all, for the course of two billion years was the only form of life on Earth, and, as the catalysts of biogeochemical cycles, formed the biosphere. Today they remain the primary biogeochemical engine of the planet. But while at one time the ancient prokaryotic bacterial communities reigned supreme, covering most of the Earth in a solid membrane as a kind of monopolistic power in the biosphere, over the course of evolution its autocatalytic units "migrated" and found themselves joined to more complex organisms, forming specialized organelles in nuclear cells—*mitochondria and chloroplasts*. Management of Gaia's "physiological" processes (restorative-oxidizing, binding oxygen to carbon, etc.) is conducted by both the direct heirs to these nucleus-free single cells such as soil bacteria, and their descendants in nuclear cells—*mitochondria* (oxydizers) and chloroplasts (deoxidizers). And this catalytic hypercycle, to use a term from Manfred Eigen, binds the smallest living organisms to the planetary macrosystem as part of maintaining the climactic and biogeochemical parameters of its environment (Eigen and Schuster 1979).

It's hard not to notice the striking similarity between Gaia and the modern representation of the biosphere in the vein of Vernadsky's ideas, of whose works Lovelock learned only in the 1980s (due to a lack of adequate translations of "The Biosphere" into English as well as, by his own admission, a "deafness" of anglophone writers to foreign languages). There are some distinctions, however. First of all, generally speaking, Gaia is not the biosphere but the Earth as a whole. Here Lovelock draws a picturesque comparison between Gaia and the cross-section of an old tree, where the living part (the biosphere) is only a thin layer of vascular tissue under the bark, and the main mass of dead timber is the product of extended activity by this layer. Second, the Gaia hypothesis takes a skeptical attitude toward the possibility of humans conquering nature and submitting it to their interests, in opposition to Vernadsky's postion.

But is it even possible to consider the "Gaia" concept, which Lovelock himself calls a hypothesis, science in the full sense of the word? And in this hypothesis, aside from grandiosely bold ideas and philosophical underpinnings, a more strictly scientific component? Here it's worth noting that several of Lovelock's "geo-physiological" hypotheses have received confirmation through scientific experimentation.

In 1981, Lovelock postulated that the global climate stabilizes itself by way of the carbon dioxide cycle's self-regulation through biogenic intensifications of the rock erosion process. In the terms of geo-physiology, carbon dioxide is a key meta-bolic gas of Gaia, influencing not only the climate, but also plant production, as well as production of oxygen in the atmosphere. The main abiotic source for this comes from volcanic activity. Carbon dioxide gas dissolved into rain and ground matter creates carbonic acid, which interacts with silicates and bicarbonates in a rock, resulting in the creation of bicarbonate ions (chemical erosion). The products of this interaction are carried off by streams to the World Ocean where plankton and coral use them to build their skeletons. After death, these tumble to the bottom of the ocean, forming a chalky residue.

The results of research by David Schwartzman and Tyler Volk, published in *Nature*, confirmed that micro-organisms and planets are able to speed the chemical erosion of rock by tens or hundreds of times (Schwartzman and Volk 1989). Also, the plants that swallow carbon dioxide from the air and transfer the carbon content into soil raise its local concentration by 10–40 times. The main mass of dead plants, undergoing bacterial oxidation, also turns into carbon dioxide at point of contact with calcium compounds, silicates and water. Thus, the biota, influencing the con-centration of atmospheric CO_2, a greenhouse gas, participates in regulating the tem-perature setting of Earth.

One could produce other examples, proven today, of a closed chain of cyclical causation, the typical characteristic of geo-physiology (Gaia theory). Lovelock's central postulate with its idea of Gaia as a global correlated superorganism, how-ever, does more poorly, having met with stern criticism from many famous evolu-tionary biologists (Ford Doolittle, Richard Dawkins, etc.). After all, the evolution of the biosphere according to the "Gaia" Concept is interpreted as the individual development (epigenesis) and improvement of its self-regulating properties. However, from the point of view of traditional scientific representations, strictly correlated and high-complexity systems (including Gaia) inevitably degrade and pull apart with time. Living organisms are also distinguished by highly complex organization, but this complexity and order is supported in nature by using a mech-anism of competitive interaction by individuals. Those who have lost internal order and, as a result, become uncompetitive, are weeded out of the population. It is through this process of evolution that the unique complexity of living materials is reproduced and supported.

But Gaia exists in the singular, and therefore cannot reproduce. Thus, Dawkins notes, a natural selection of the most adaptive planets is impossible. And, therefore, there can be no discussion of any extended preservation of Gaia's self-regulating abilities without the ordering will of a Creator standing behind her. Or, Dawkins notes sarcastically, a committee of species that assembles annually for the purpose of deciding the climate and chemical makeup of the planet for the following year. Lovelock couldn't come up with anything to oppose this criticism, and the scientific community recognized the untenability of the theory as a whole (despite its undeni-able beauty).

Further on we will tell of how St. Petersburg biophysicist Viktor Gorshkov attempted to resolve this problem. But now we will return to the already cited work of Nikolay Timofeyev-Resovsky, in which he, even before Lovelock, was able to find an approach to overcoming this contradiction. He called attention to the structural unit of the biosphere, within which the natural selection of populations occurs. These are biocenoses, he says, "the elementary units of the biological cycle, i.e. of the biogeochemical work taking place in the biosphere."

Timofeyev-Resovsky continues, "The majority of biocenoses are in a state of prolonged dynamic equilibrium, being very complex self-regulating systems. So the problem of studying the causes, mechanisms and support conditions for such a dynamic equilibrium in biocenoses is especially important." And without knowledge of these mechanisms, "it is impossible to understand and properly schematize the true occurrence of evolutionary processes in nature, constantly improving in dynamic biocenoses and their greater complexes—landscapes" (Timofeyev-Resovsky 1996, p. 63).

It's not hard to note how different this structured system of "biospheric cells" is from the concept of "Gaia." After all, if the work of supporting the biogeochemical cycle is performed not by the biota overall, or by some anthropomorphized "super-organism," but by separate biotic communities and their populations, it therefore leaves room for competitive interaction. That is the mechanism for weeding-out and replacing poorly working "cells" which protects the biosphere from degradation and collapse, preserving its capacity to support global biogeochemical balance for an indefinitely long period of time. But we will speak in more depth of this in the following chapter, in connection with Victor Gorshkov's concept of biotic regulation of the environment. For now, let us again conduct a mental overview of the path ecology has taken from the moment of its establishment as an independent branch of science.

When, in the late 1920s, Vernadsky came to the idea of the biosphere as a single holistic entity forming the face of our planet, and Tansley soon after introduced ecology's key understanding of the ecosystem, the majority of people still imagined the world to be open and nearly limitless, a place where man could do whatever he saw fit, and could adapt and remake according to his needs. What ecologists did within laboratory walls seemed far away from people's everyday business and worry. It would take more than half a century to make the connection obvious and to make terms like biosphere and ecosystem equal in usage to understandings such as energy and evolution. Nonetheless, that path has not yet come to its end. Between acknowledging human dependence upon the environment and understanding the full danger of its degradation lurking in the none-too-distant future as ecologists warn stands an enormous distance. But cross it we must, if the future is to come at all.

Table 9.1 Scientists who have made contributions to the formation of a systemic concept of the biosphere

Haeckel's line	Humboldt-Vernadsky line
Ernst Haeckel (1834–1919)—German evolutionary biologist, follower of Charles Darwin, first introduced the concept of ecology as an area of biology, studied interaction of organisms with environment.	Edward Suess (1831–1914)—Australian geologist, first used the term biosphere in the sense of one of Earth's coverings, alongside the atmosphere, hydrosphere and lithosphere.
Karl Mobius (1825–1908)—German zoologist and hydro-biologist. Using the example of oyster beds on the North Sea, developed and proved theory of an internally created community of organisms populating one or another similar area of the sea floor, which he called biocenosis (1877).	Vladimir Vernadsky (1863–1945)—Russian mineralogist and geochemist. Put forward and developed concept of the biosphere as a holistic and interconnected world of living material, united by a system of biogeochemical cycles with the abiotic spheres—atmosphere, hydrosphere, and lithosphere. Showed that the chemical state of our planet's crust is in large part under the influence of life and is determined by living organisms.
Henry Cowles (1869–1939)—American botanist. Studying vegetation on the shores of Lake Michigan, discovered regular stages in development in different types of ecosystems and first described the process of biological succession.	Alfred Lotka (1880–1949)—American mathematician, author of the book *Elements of physical Biology* wherein he first made attempt to remake biology as a strictly quantitative science. Developed mathematical models for interspecies interaction (for example, a model describing the interconnected trends in numbers of predators and prey) as well as biogeochemical cycles.
Frederick Clements (1874–1945)—American botanist, developed concept of succession in detail. Introduced concept of climax communities as the ultimate stage of biological succession, showed adaptiveness of biocenoses, their ability to adapt and evolve as the environment changes.	Nikolay Timofeyev-Resovsky (1900–1981)—Russian biologist, one of the founders of molecular genetics and radiobiology. In later life focused on global problems of biology. Compared biosphere to giant living factory forming Earth's environment. First pointed to the role of biocenoses as elementary cells in biological cycle and to the possibility of competitive relations between them, creating conditions for stabilizing evolutionary selection.
Charles Elton (1900–1991)—English zoologist. Laid groundwork for population ecology. Introduced understanding of ecological niches and formulated rules for ecological pyramid-reduced numbers of organisms from lower trophic levels to higher ones.	James Lovelock (born 1919)—English scientist, electrical engineer by education. Put forward "Gaia Hypothesis," an original concept of Earth as a holistic superorganism in which the evolution of living things is closely linked to changes in their physical and chemical surroundings. This concept enabled a new way of thinking about global mineral cycle processes. Many of Lovelock's theoretical predictions were confirmed through experimentation. However, Lovelock was unable to explain how this superorganism, a complex system of correlations, has avoided inevitable degradation and collapse for hundreds of millions of years, for which he suffered criticism from evolutionary biologists.
Arthur Tansley (1871–1955)—English botanist. Introduced concept of ecosystem in 1935, considered founding year for ecology as independent branch of science. Tansley's main accomplishment was the successful attempt to integrate biocenose and biotope at level of new functioning unit—the ecosystem, which became for ecology what the atom is for physics, the molecule for chemistry or the cell for cellular biology.	
Vladimir Sukachyov (1880–1967)—Russian biologist, forest specialist. Using example of forest communities, developed concept of the biogeocenose (1940), analogous to the ecosystem, distinguished by limited scale and uniformity, natural and climatic conditions and including biotic and abiotic components of the environment.	

(continued)

Table 9.1 (continued)

Haeckel's line	Humboldt-Vernadsky line
Georgy Gause (1910–1986)—Russian biologist, modeled processes of interspecies competition in bacterial and protozoan cultures. Formulated the competitive exclusion principle, according to which, two species cannot occupy the same ecological niche: one of the species inevitably pushes out the other.	Viktor Gorshkov (born 1935)—Russian theoretical physicist. Developed concept of biotic regulation of the environment. Unlike Lovelock, Gorshkov linked the problem of supporting environmental parameters beneficial to life with the life activities of competing, independent biotic communities (biocenoses).
Raymond Lindeman (1915–1942)—American ecologist. Developed general scheme for transformation of energy in an ecosystem and the basic methods for calculating its energy balance. Demonstrated the rule of trophic pyramids: As energy transfers from one trophic level to the next, the quantity reduces by an order of magnitude. Thus, an organism of a higher level has access to no more than ten percent the energy used by an organism of the previous level.	Reacting to disruption of the environment with system change or destruction of organic material, the biota is able to absorb excesses of one or another nutrient in the environment or, contrarily, fill a deficit of it and thus regulate concentrations at a level suitable for life. According to Gorshkov's concept, the current global ecological crisis arises primarily from the destruction of ecosystems over an enormous swath of land, and a transition to sustainable development is possible only if a considerable portion of destroyed ecosystems are restored.
Eugene Odum (1913–2002)—American biologist. Laid the basis for ecology as an independent scientific discipline. Known for his work in the area of ecosystem ecology, as well as the textbooks *Fundamentals of Ecology* (with Howard Odum) (1953), *Ecology* (1963) and others, which played a major role in establishing ecology as a university course. Odum brought together materials that had been scattered throughout journal articles and separate monographs, reassembling them into an omnibus of basic concepts for ecology.	
G. David Tilman (born 1949). American ecologist. A notable representative of a new branch of ecology focused on physiological mechanisms. Tilman's most famous works are in the research area of limiting resources based on diatom algae and grasses. Proved that ecologically similar species can coexist if they are limited by different resources (for example, the concentrations of various nutrients dissolved in water).	

References

Eigen, M., & Schuster, P. (1979). *The hypercycle: A principle of natural self-organization*. Berlin: Springer.

Elton, C. S. (1946). *The ecology of animals*. London: Methuen.

Lindeman, R. L. (1942). The trophic-dynamic aspect of ecology. *Ecology, 23*, 399–418.

Lotka, A. J. (1925). *Elements of physical biology* (p. 460). Baltimore: Williams & Wilkins.

Lovelock, J. E. (1991). *Gaia: The practical science of planetary medicine*. London: Gaia Book Limited.

Odum, E. (1983). *Basic ecology*. Philadelphia: W.B. Saunders.

Reymers, N. F. (1990). *Prirodopolzovanie. Dictionary-reference*. Moscow: Mysl. [in Russian].

Schwartzman, D. W., & Volk, T. (1989). Biotic enhancement of weathering and the habitability of earth. *Nature, 340*, 457–460.

Timofeyev-Resovsky, N. V. (1996). The biosphere and humanity. In A. N. Tyuryukanov & V. M. Fedorov *N. V. Timofeyev-Resovsky; Thoughts on the biosphere. Thoughts on the biosphere*. Moscow: RAEN. [in Russian].

Vernadsky, V. I. (1988). *Philoshophical thoughts of a naturalist* (p. 519). Moscow: Nauka. [in Russian].

Vernadsky, V. (1998). *The biosphere: Complete annotated edition, translated by D. Langmuir*. New York: Springer-Verlag/Copernicus Books.

Zavarzin, G. A. (2004). *Lectures on microbiology in the context of natural history*. Moscow: Nauka. [in Russian].

Part IV
Permanence of the Planetary Environment and the Concept of Biotic Regulation

Chapter 10
Abiotic Factors in Forming the Earth's Climate

Despite all of the weaknesses of the Gaia Hypothesis and its just criticism from evolutionary biologists, Lovelock's ideas have borne an unquestionable revolutionary influence upon the minds of researchers. Meanwhile, the ecological situation itself in the last quarter of the twentieth century—the progressive reduction in the area of land ecosystems, increased concentrations of greenhouse gasses, the warming climate, the wiping out of tropic forests, the expansion of deserts and semi-deserts—has spurred scientists to switch their attention from the problems of environmental pollution to the processes of ecosystem destruction. Even economists have begun calling them the foundation of life. Humans are changing the biosphere faster than they can understand it, as was said in the article "Human Domination of Earth's Ecosystems" published in the journal *Science* by a group of high-profile American ecologists, and we must redouble our efforts to study it while it is still possible to bring the processes destroying the biosphere under control (Vitousek et al. 1997).

In this way, ecology has come to understand the objective need for deeper consideration of biospheric processes which would allow us to overcome the narrow approach to the assessing of the global crisis typical of the first reports to the Club of Rome. From the other side, this consideration must also innovatively illuminate the strategic role of the biota as shown by data acquired across life and earth sciences. One such attempt, developed by St. Petersburg biophysicist Viktor Gorshkov in the mid-1990s, was the theory of biotic regulation of the environment, which we will pause upon in this section of the book.

Unlike Lovelock, Viktor Gorshkov never worked at NASA or had any relationship with the Soviet Space Program, but "a look from the cosmos," allowing the extrapolation of conditions on Earth and its neighboring planets, played no small role in formulating his concept. The impressive sustainability of Earth's biota served as grounds for the comparison, demonstrated over billions of years of existence. It was this very sustainability that the Gaia Hypothesis proved unable to explain.

Truly, how, on this planet, on this (to borrow Lev Itelson's phrase) "unbelievably weak, unsustainable flame, wind-swept and set adrift by the universe," despite all

© Springer International Publishing AG, part of Springer Nature 2018
V. I. Danilov-Danil'yan, I. E. Reyf, *The Biosphere and Civilization: In the Throes of a Global Crisis*, https://doi.org/10.1007/978-3-319-67193-2_10

geological and climatic perturbations suffered over a cosmic timescale, did this feeble torch of life never once, it seems, go out? How, in spite of all the cataclysms—collisions of asteroids with the Earth, a cooling climate and glaciation, gargantuan volcanic eruptions, shifts and cracks in the Earth's crust, spikes and troughs in the level of the World Ocean—did the biota ultimately come out the victor? And what on earth is supporting life on this thin coating, to put it neatly, between the blistering mantle and the eternal cold of interplanetary space and its deadly cosmic radiation? Does it owe its existence only to a conjunction of physical conditions, including the uniquely fortunate near-sun positioning of the Earth's orbit, or is that just a favorable stage for the unfolding of other, non-coincidental processes and phenomena supporting the conditions necessary for life?

Indeed, the existence of life on Earth is possible within the relatively narrow bounds of temperature where water is in its liquid phase. Even a few degrees below freezing is already the temperature extreme for the vast majority of species, and only a few warm-blooded creatures can long exist actively at temperatures below zero Celsius (Emperor Penguins can even reproduce). At temperatures above 60 °C (140 °F), only a small number of thermophilic bacteria maintain viability.[1] The optimal temperature range of the great mass of organisms lies somewhere between +10 and +20 °C (50 and 70 °F). And, as radioisotopic research of rock and sediments has shown, that is the range the average surface temperatures on our planet has kept for the past 600 million years, dropping to +10 °C during periods of glaciation and rising to +20 °C at times of maximum warming. So, how do we explain this impressive sustainability in temperature and climate?

Some of the reasons are quite obvious and overall well known to modern science. Most of all, this is the stability of solar radiation and consistency of the sun's light energy reaching the Earth—roughly 174×10^{15} W. Of this, about 30% is immediately reflected by clouds, atmospheric dust or aerosols and ice or snow-covered sections of Earth's surface without any warming effect. Another 23% remains in the atmosphere spent on water evaporation. Only 47% reaching the Earth itself, warms the surface of land and sea, and from this, in turn, through secondary (long-wavelength) infrared radiation, the atmosphere warms (Marfenin 2006, p. 58). No less a material role is played by the short Earth day (the changing of day and night), almost circular orbit of Earth, and also that its axis tilts at an incline of about 66° in its elliptical orbit, which allows more or less even warming of the planet's surface over the course of the seasons.

An enormous role in maintaining the Earth's climate and temperature regime is played by the powerful atmosphere and hydrosphere, capable of accumulating an enormous amount of heat energy, and the Earth's mass (6×10^{23} metric tons) allows it to hold this powerful layer of water and air. To give some idea of the gigantic quantity of heat accumulated by the World Ocean, let us recall that it occupies 70.8% of the planet's surface and contains 1320–1380 km³ of water, whose unit thermal capacity is equal to 1 cal/g to 1 °C or 4186 J/kg to 1 °K (the highest thermal

---

[1] On the ocean floor, in hot sulfur springs, bacteria have been found that can survive under conditions of high pressure even at a temperature of +115 °C.

capacity of all known liquids). Due to this, stores of heat in the World Ocean surpass the yearly quantity of thermal energy the Earth receives from the Sun 21 times over. So we would be entirely justified in calling the hydrosphere the most important temperature stabilizer on the surface of our planet (Marfenin 2006, pp. 26–29).

The alignment of the climatic heating regime also enables constant motion of air masses in the lower atmosphere and powerful ocean currents from the equator to the poles and back, creating, as a result, conditions acceptable to life at all latitudes. Along with this, the atmosphere is constantly filling with moisture evaporated from the surface of the World Ocean, adding up to 500,000 km^3 or 86% of all water evaporated from the Earth's surface. An enormous quantity of energy (660 cal/g) enters the atmosphere in the process of evaporation, which is carried through air flows. Winds, cyclones and hurricanes are born of this global energy redistribution machine. In just the same way, water vapor, before pouring as rain or falling as snow, can be carried by air flows for thousands of miles. About 90% of water, evaporated from the World Ocean, settles back upon its surface creating the greater or oceanic water cycle. The other 10% precipitation, coming down on land, makes up the lesser or continental water cycle. Most of the fallen water then returns with river drainage to the World Ocean, though some part of it is held back in glaciers (Marfenin 2006, p. 56).

The presence of water vapor in the atmosphere enables it to retain part of the heat energy that the planet radiates---the greenhouse effect. It is estimated that the greenhouse effect adds about 30 °C to average surface temperatures. Of that, water vapor makes up a 20.6 °C share. The second most significant greenhouse gas, carbon dioxide, contributes 7.2 °C. After that comes ozone (2.4 °C), nitrous oxide (1.4 °C), methane (0.8 °C) and other, less active greenhouse gasses.

But while concentrations of water vapor in the atmosphere, depending on altitude, geographic latitude and season, add up to between 0.5 and 4% or about 1% on average, the total concentrations of CO_2 and other greenhouse gasses never exceed a tenth of a percent. Nonetheless, even these seemingly insignificant admixtures turn out to be enough to prevent the Earth from freezing.

It's not hard to see, however, that the greenhouse effect has another side, and even a small increase in the atmospheric concentrations of greenhouse gasses is capable of increasing the average surface temperature. And this, due to a highly complex climate system, is fraught with the most unpredictable consequences. Thus, for example, when the temperature of the World Ocean's surface layer rises, the solubility of carbon dioxide gas in the water decreases, leading to disruption of the buffer equilibrium between ocean and atmosphere, raising the level of atmospheric CO_2 and, consequently, intensifying the greenhouse effect. Beyond this, the process of thawing glaciers and permafrost is accompanied by the release of constituent CO_2 and methane, as well as the buildup of water vapor in the atmosphere. In this way, there arises a kind of (intensifying) positive feedback loop with the effect of accelerating warming of the climate. For this and a number of other reasons, specialists today are very worried about the danger of destabilizing the Earth's climate system, demanding the adoption of drastic measures to limit anthropogenic impact on the environment.

***

If the World Ocean serves as an accumulator for heat on the planet, then glaciers, both sheet (Antarctica, Greenland) and mountain, could justifiably be called accumulators of cold, whose share in forming the Earth's climate and stabilizing surface water stocks on land is hard to overestimate. For example, over two-thirds of world fresh water reserves are concentrated in glaciers. Their total runoff to the ocean, 3450 km³ of water, makes up roughly 8% of total surface water runoff. They are also responsible for 3% of rainfall on land. Thus playing the role of natural reservoir, glaciers enable the redistribution of drainage from atmospheric rainfall over the course of the year, which is especially important for regions where they feed the rivers a significant part of their flow, such as the Yukon (23% glacial runoff), Kuban (6%), the Indus (8%), the Syr Darya (6.5%) and the Amu Darya (15%). In Central Asia, where glaciers occupy a mere 5% of territory, the share of glacial runoff in rivers over summer months reaches 50%.

At the same time, glaciers aid the cooling of the atmosphere by increasing the Earth's albedo,[2] as well as expending heat on ice melt and compensating glacial radiation (the cooling of air above their surface). Occupying 3% of the planet's area, glaciers reflect 5% of solar radiation into space, which raises yearly average albedo from 0.29 to 0.3 and cools the surface layer of air by roughly 1 °C. Furthermore, the albedo mechanism for cooling the atmosphere is supplemented by the outflow of turbulent heat from the atmosphere into glaciers, of which 6% is expended on ice melt and 94% on radiating it, mainly in Antarctica. The total sum cooling of the Earth's surface by modern glaciation makes up about 2 °C (Govorushko 2006, pp. 60–70).

In this way, while the World Ocean has for all time aided in the conservation of heat on the planet, the mission of glaciers has directly opposed it, and the interrelation of these forces in different geological epochs has been far from uniform. Science knows of no few episodes in the history of Earth when the average surface temperature dropped by 5–6° or more, and glaciers covered enormous swaths of land. At times, this shell of ice crushed whole continents at up to two kilometers thick, and sea level, compared to the present, dropped 100–120 m. The last such glaciation took place 252–12,000 years ago, and the whole Quaternary period of geologic time beginning 1.8–1.6 million years ago, which includes the present Holocene epoch, due to repeated, powerful glaciations, received the name of Ice Age.

The causes of these periodic glaciations are still not entirely clear. In all likelihood, they are the result of several factors, one of which plays a leading role, the others going into action as part of a launch mechanism. One notices, for example, that all great glaciations on our planet coincide with the most massive mountain-building epochs, when the Earth's surface relief was at highest contrast and the sea area correspondingly decreased. Under these circumstances, the climate varied more sharply. Thus, the mountains of up to 2000 m (6562 ft) in height, arising 30 million years ago in Antarctica, became the epicenter for the formation of sheet

---

[2]Albedo—the reflective capability of the Earth's surface or that of other celestial bodies, more exactly—the relative value of a flow of reflected (scattered) radiation to the flow of incoming radiation.

glaciers, and the continent itself—a giant accumulator of cold on the planet. At the same time, the formation of the enormous Tibetan Plateau made a material contribution to climate change in the northern hemisphere, along with events in western North America, where conditions arose for the chemical erosion of rock and accelerated removal of carbon dioxide, the planet's heat switch, from the atmosphere. All of this together enabled a gradual cooling of the climate and ultimately led, three million years ago, to periodic glaciation of most of the northern hemisphere (the Quaternary, or Ice Age, Period) (Rezanov 1984).

Volcanologists in Kamchatka managed to trace the connection between glaciation and volcanic activity, whose role was greatly underestimated until recent times. Volcanic eruptions, as you know, are capable of noticeably influencing the Earth's atmosphere, changing its gas composition and fouling the air with aerosols and volcanic ash, causing a decrease in transmission for the visible part of the solar spectrum. Particularly powerful eruptions may be accompanied by a reduction in solar radiation by 10–20%. Furthermore, particles emitted by volcanos serve as a nucleus of condensation, aiding the development of cloud cover. According to estimates, an increase in cloud cover from 50% (typical for our time) to 60% is able to reduce the Earth's average surface temperature by 2 °C (Melekestsev 1969, pp. 140–149).

Finally, it's worth naming one more possible cause of the great glaciations that have a purely cosmic origin. According to a version put forward by Serbian mathematician and geophysicist Milutin Milankovic (1879–1958), most of the known hot and cold cycles in the climate are connected to periodic changes in the Earth's position in the solar system (thus called Milankovic Cycles). These are variations in the tilt of the Earth's equatorial plane toward the ecliptic plane from 21.5 to 24.5°, completed over the period of 41,000 years, as well as the rotation of the Earth's axis over a 23-thousand-year period. A change in form to the Earth's orbit exerts a weaker influence, periodically rounding or elongating over the course of about 100,000 years. Put together, all these factors may reduce sunlight penetration into the northern hemisphere by 20% (Imbrie and Imbrie 2005).

***

And so, despite a powerful atmospheric and oceanic buffer, the Earth's climate system is in a state of unsustainable equilibrium. On the one side, the greenhouse effect lies in wait, ready to spring as the result of any chance fluctuation that brings with it an increase in average surface temperature. The causes might be varied, such as major emissions of carbon dioxide into the atmosphere linked to volcanic activity. In any case, even a small warming of the World Ocean intensifies evaporation from its surface and a growing concentration of water vapor in the atmosphere. Further, this initiates a positive feedback loop: A growing greenhouse effect enables still greater warming of the Planet's surface and, therefore, ongoing accumulation of atmospheric moisture, and it's hard to predict where the process might end. On Earth's neighbor, Venus, for example, it has led to the atmosphere heating up to +475 °C (887 °F) and the total evaporation of the oceans.

On the other hand, a lowering of the Earth's average surface temperature presents the danger of the opposite effect—decreased evaporation and a weakening of the greenhouse effect. This process, however, cannot end with the total disappearance

Fig. 10.1 Schematic illustration of physical unsustainability in Earth's climate (from V. G. Gorshov and A. M. Makarieva http://www.bioticregulation.ru/life/life2_r-3.php)

of water vapor from the atmosphere: when the Earth's average surface temperature reaches negative figures, the initiating glaciation of the planet leads to an increase in its reflective capability, which means further cooling as well. The situation on Mars, in part, bespeaks of such a situation. As such, these two extreme states are truly sustainable, or you might say, physically assigned. And one of the most important factors maintaining a stable climate on Earth, according to Gorshkov, is the biota which has settled it, i.e. life itself, which, plugged into the biogeochemical cycle works to create climatic conditions favorable to itself. For example, absorbing excess carbon dioxide and thus removing it from the atmosphere. Or the opposite, releasing it in the process of organic decay, increasing the atmospheric concentrations of CO_2 required to support the temperature regime necessary for life.

In order to prove this hypothesis, Gorshkov attempted to calculate the balance of Earth's climate, based on Alexandr Lyapunov's sustainability principle and putting the Earth's biota in the parentheses. Not having the chance to pause upon the mathematical side of these calculations, we'll just say that when he plotted the graph, the curve for Earth's current temperature and climate parameters was "latched" to the very top, i.e. to the zone of high unsustainability. Even the smallest incline away from the average surface temperature one way or another inevitably led to irreversible climate change (Gorshkov et al. 1999; Makarieva and Gorshkov 2001).

All of this can be schematically illustrated by likening the world's climate to a ball on the peak of a little hill (Fig. 10.1). The ball is able to balance itself on this tottering position for some time, but only the slightest outside pressure is enough for it to lose balance and roll down to the foot. Therefore, a position at the foot of the hill is truly sustainable for the ball, or physically determined.

But what if we carve a hollow in the shape of our ball into the peak? Then we have created a physically determined position for it, in which it can remain for an indefinitely long time. Just such a thing, Gorshkov says, is taking place in the case of the

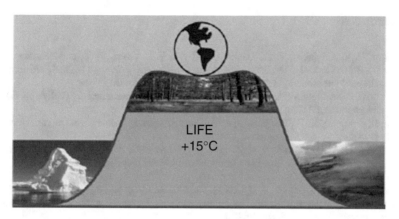

Fig. 10.2 Biotic sustainability of the Earth's climate (from V. G. Gorshov and A. M. Makarieva http://www.bioticregulation.ru/life/life2_r-4.php with changes)

Earth's climate. Only it is the biota itself that creates this "dimple of sustainability," providing it not a physically, but a biotically determined sustainability (Fig. 10.2).

Truly, as data from Palaeothermometric research testifies, the earth has never witnessed serious climate shifts to one side or the other in the course of geologic time, including periods of great glaciation, when ice sheets bound enormous swaths of land like armor, descending to the latitudes of New York in America and Frankfurt-am-Main in Germany (Monin and Shishkov 1979; Koronovsky and Yakushova 1991). Thus, more than once the Earth's climate has undergone a very severe trial. And, nonetheless, the "climate ball" remains within bounds of parameters acceptable to life, which is hard to explain without accounting for the global role of the biota, which is clearly not only a result, but also the cause, or at least one of the causes of temperature and climate conditions favorable to it.

And then before the researcher inevitably arose the question of the mechanism for this stabilizing influence. The conclusion of Gorshkov's scientific quest was his monograph *Physical and Biological Bases for Sustainable Life* (1995), in which he undertook an attempt to link together an understanding of global biogeochemical processes, their role and place in the world's biota and the destructive potential that human economic activity represents. We don't have the chance to reproduce his scientific argument here in its full scale, so anyone who would like to more fully acquaint themselves with the concept of biotic regulation of the environment may refer to Vikor Gorshkov and Anastasia Makarieva's website: http://www.bioticregulation.ru/index.php We will try to elucidate here its key point and acquaint the reader with the most important conclusions that follow from it, relating to the problem of sustainable development at the same time without avoiding its controversial and speculative parts.

References

Gorshkov, V. V., Gorshkov, V. G., Danilov-Danil'yan, V. I., Losev, K. S., & Makarieva, A. M. (1999). Biotic Control of the Environment. *Russian Journal of Ecology, 30*(2) 87–96. Retrieved from http://www.bioticregulation.ru/common/pdf/ecol99-en.pdf.

Govorushko, S. M. (2006). *Glaciers and their importance for human activity* (Vol. 6, pp. 60–70). Vladivostok: Pacific Geographical Institute [in Russian].

Imbrie, J., & Imbrie, K. P. (2005). *Ice ages: Solving the mystery.* Cambridge: Harvard University Press.

Koronovsky, N. V., & Yakushova, A. F. (1991). *Fundamentals of geology.* Moscow: Vysshaya shkola. Retrieved from http://geo.web.ru/db/msg.html?mid=1163814&uri=part22-07-1.htm [in Russian].

Makarieva, A.M., & Gorshkov, V. G. (2001). Greenhouse effect and the problem of stability of the average global temperature of the Earth's surface. *The reports of the Russian Academy of Sciences, 376*(6). 810–814. [in Russian].

Marfenin, N. N. (2006). *The sustainable development of humanity* (p. 624). Moscow: MGU. [in Russian].

Melekestsev, I. V. (1969). *Volcanism as a possible cause for glaciation.* In *Volcanoes and eruptions* (pp. 140–149). Moscow: Nauka. [in Russian].

Monin, A. S., & Shishkov, Y. A. (1979). *History of the climate* (p. 408). Leningrad: Gidrometeoizdat. [in Russian].

Rezanov, I. A. (1984). *Great catastrophes in the history of earth* (p. 176). Moscow: Nauka. [in Russian].

Vitousek, P. M., Mooney, H. A., Lubchenco, J., & Melillo, J. M. (1997). Human domination of earth's ecosystems. *Science, 227*(5325), 494–499.

Chapter 11
Role of the Biota in Forming the Environment

Since Vernadsky's time, we have known about the singular role that the planet's living material plays in forming the global biogeochemical cycle due to the part it takes in high-speed chemical reactions. Today there is no need to explain to anybody how life has changed the face of the Earth, and that it is to this the Earth owes its oxygenated atmosphere. The quantitative characteristics of some nutrient cycles, however, including the carbon cycle which forms the basis of organic molecules, the basis of life, became known to scientists relatively recently. That, in turn, allowed us to approach an understanding of many processes taking place in the biosphere.

So, for example, scientists established that stores of inorganic carbon available to the biosphere, unlike in ancient geological epochs, today are materially limited, adding up to about 10^3 gigatons (Gorshkov et al. 2000, p. 117). Primarily this is carbon dioxide gas dissolved in the World Ocean, as well as soil humus, peat bogs, and finally the small admixture contained in the atmosphere—less than 0.04%. However, the great majority of carbon dioxide dissolved in the ocean is located at depths below 200 m and goes barely used by phytoplankton and its emergence from the deeper strata as a result of vertical cycling is relatively low. So, the biota makes extremely economical use of inorganic carbon.

But the life strategy of ancient biotic communities in long-gone epochs looked entirely different, when the atmosphere was much richer in carbon dioxide. The colossal deposits of oil and coal, which have survived to this day and which we utilize so widely, indirectly testify to this. According to the remains of organic communities buried in ancient geological formations, which had not been reprocessed by living organisms, we can assume that, at the first stages of the Biosphere's evolution, waste from life processes emerged from the environment and stored itself in sediments (oil), or that rotting plant remains piled up faster than their bacterial decomposition occurred (coal in the Devonian and Silurian). You could characterize such a type of biotic strategy as high-entropy i.e. accompanied by significant necrosis in the organic community and a high product of mortmass (Krasilov 1992, p. 32).

Therefore, the necessity to more efficiently use the trophic resources of the environment led, over the course of evolution, to a kind of "emissions-free technology"

V. I. Danilov-Danil'yan, I. E. Reyf, *The Biosphere and Civilization: In the Throes of a Global Crisis*, https://doi.org/10.1007/978-3-319-67193-2_11

on the basis of which modern natural ecosystems function. This enables, first of all, a more complex structure of biotic communities as the diversity of life forms increased, and also the correlation of species within communities, providing the opportunity to repeatedly use the organic material created by producer organisms in the process of synthesis.

Thus arose the multi-tiered system of a more-or-less closed cycle of matter, which included a number of consecutive reprocessing stages for organic production (the food chain), starting with the autotrophic producers and ending with reducer organisms (fungi and bacteria) at the exit. The latter, reprocessing decayed organic matter to its ultimate sub-molecular compounds, makes it accessible to the root system of plants, in this way providing the opportunity to use the nutrient chemical elements necessary for life. That is, it is as though they restore the dead to life.

For the modern biota, however, adapted to an oxidized atmosphere, an oversupply of carbon is just as unacceptable as a deficit—recall the danger of an excessive greenhouse effect. By the way, the main mass of inorganic carbon is concentrated not in the atmosphere or the World Ocean, but in the earth, at a coefficient of 28,570 (lithosphere): 57 (ocean):1 (atmosphere) (Marfenin 2006, p. 80). That is even with the regular release of large quantities of inorganic carbon from the earth into the hydrosphere and atmosphere by way of volcanic activity, degasification of magma and rifts in the ocean floor. Volcanic gasses contain a 15% share of carbon dioxide, 75–80% of water vapor and of all other gasses (CO, CH_4, NH_3, H_2S, etc.) a total share of no more than 10%. However, in the Earth's distant past, there were periods of much greater volcanic activity compared to today, and nonetheless even against a backdrop of major variations, atmospheric concentrations of CO_2 maintained their order of magnitude over the course of hundreds of millions of years (Broecker et al. 1985; Barnola et al. 1991).

The fate of carbon emitted into the atmosphere is already no secret today. In past geological epochs, it was all reliably buried in sedimentary rock—the remains of the fossilized biota in long-gone biosphere (chalk, limestone), as well as varieties of hydrocarbon fuel (oil, coal, peat, shale oil, natural gas, etc.). All of this is linked to the activity of the biosphere, removing superfluous carbon from circulation over many thousands of years. The scale of the stream removed is assessed by experimental means, thanks to the discovery of tiny granules of organic carbon, kerogen, sown in geological deposits. From this, it became clear that carbon was stored at a rate of 0.01 gigatons per year over the past 600 million years (Budyko et al. 1987).

A special place in the regulation of the global carbon cycle belongs to the biota of the World Ocean. 57 times more carbon dioxide gas has dissolved itself in the ocean than is contained in the atmosphere—40,000 gigatons of carbon to 700 (Bolin 1983). And there is a reason why it does not diffuse into the Earth's atmosphere— the carbon-dioxide poor layer of water under the ocean surface.

This layer between 100 and 200 m in depth is called the photic zone, as this is the distance below the surface that rays of light on the visible part of the solar spectrum penetrate, thus establishing conditions for photosynthesis. It is warmer than the deep, practically unheated layers of the ocean. It floats, you might say, above the cold depths, separated by a zone of sharp, bounding changes in temperature and pressure called the *thermocline*. Aside from this, the near-surface layer is well stirred by the wind, which, combined with the chemical properties of water-dissolved carbonates provides for atmospheric CO_2 to be swallowed up quickly enough and its concentrations in air and water balanced according to Henry's Law.

The main particularity of the photic zone, however, is the presence of phytoplankton, which reworks the great mass of waterborne organic material in the process of photosynthesis and serves as food for all ocean consumers. Photosynthesis, as you know, involves the absorption (bonding) of carbon dioxide. The volume of this abortion in the World Ocean is estimated to be at the order of 40 gigatons of carbon per year (Green et al. 1984, vol. 1, Ch. 9.2.1).

In this way, the photic zone plays the role of a unique buffer, swallowing up carbon dioxide gas accumulated in the atmosphere on one hand, and preventing excess CO_2 dissolved in the ocean from entering the atmosphere on the other. This last mechanism received the name of biotic pump, since CO_2 welling up from the sea depths as a result of storms and upwelling[1] enters a process of biotic synthesis at the surface. Afterward, as the organisms living there die off, this bonded carbon again sinks to the ocean floor. There it accumulates in the form of dissolved organic substances or as calcium carbonate ($CaCO_3$) in organisms' limestone skeletons and is buried in dense deposits, then partially released in the process of decomposition.

By the way, as geologist Nikolay Koronovsky writes, "Until entirely recently, this role of biogenic deposit accumulation was still clearly underestimated. Now it is established that of the whole mass of deposits, biogenic material accounts for 50–60% and each year greater than 350 billion tons accumulate in conversion to solid matter. The material dissolved in the water is digested by the aquatic biota that filters ocean water. The sea biota requires only half a year to filter through itself all the water in the World Ocean" (Koronovsky 2003). Thus, the ocean biota carries its weight to maintain the composition of seawater it needs, preserving it practically intact over the course of the whole Phanerozoic Eon, or 600 million years. Analogously, the same mechanism enables the removal of excess atmospheric carbon, absorbing it through a process of organic synthesis and partially burying it on the ocean floor.[2]

If we could transport a modern person in H. G. Wells' time machine back a billion years and give them a chance to look at dry land from a birds-eye view, they would not only fail to see any trace of vegetation, but would not even encounter any landscape as we typically use the word. Instead of rivers flowing along their channels, there would be some kind of limitless delta with countless ducts and rivulets. Instead of a defined shoreline dividing land and sea—a half-flooded space of many

---

[1] Upwelling—ascension of ocean water from its deeper strata under the influence of wind forcing warm water from the surface or a number of other causes. The particular importance of this process is linked to the welling up of various nutritional components from the depth, enriching the surface strata and increasing its bio-productivity.

[2] Here we will not touch on the details of nutrient cycles for nitrogen, phosphorus, sulfur, or other so-called minor nutrients, the share of which in the general mass of organic material is relatively small, adding up to no more than 1%.

miles, neither land nor sea. Kirill Eskov wrote, "There are serious grounds to suppose that continental landscapes of the modern look were not there at all" (Eskov 2000). Paleontologist Alexandr Ponomarenko thought (in 1993), "The existence of true bodies of fresh water, whether flowing or still, was very problematic until vascular plants somewhat reduced the speed of erosion and stabilized the coastline." There is even the opinion that in these distant times, land at a certain distance from the shoreline was deprived of any moisture whatsoever, since all rainfall had to either fall over the ocean or nearby (Gorshkov and Makarieva 2007).

In other words, living organisms not only made landfall, but in some sense created land as we know it, and the decisive contribution in establishing continental landscapes of the modern look was made by higher (vascular) plants. The key moment came with the biota's soil formation function.

But how could soil appear upon this lifeless rock with its irregular water supply? As Vasily Dokuchaev (1846–1903), one of the founders of soil science, showed in one of his works, soil represents a very complex formation—a natural body of organic material and minerals arising as a result of influence from the biota and physio-chemical factors upon continental rock. One such factor, the process of destroying and eroding mountain rock, was until recent times thought to occur by the actions of sun, wind, falling temperatures, freezing and thawing of water falling into cracks. Only in the past decades have we managed to reveal the enormous role in this process of living organisms, primarily bacteria and fungi, which accelerated erosion 100–300 times.

Landing on eroded rock, they dilute and destroy its surface layer, where, after they die, hollows form, holes and fissures filled with the dry biomass of fallen organisms. Mosses settling on rock surfaces thus prepared draw from them chemical elements necessary for life, and also aid in creating organic acids, sharply accelerating the dilution and hydrolysis of minerals. The biota becomes an active supplier of debris, able, as it accumulates, to hold moisture together with its diluted organic and inorganic compounds, thereafter serving as a fitting substance for plant seeds to grow in. And their growing roots, branching out and penetrating that newly developed layer of soil, assist its structuring and sturdiness, i.e. the formation of our familiar landscape forms.

From the above, it is clear how great a role the biota plays in soil formation, despite the relatively small share of organic material in the soil (about 10%). But the incessant cycle of soil elements takes its properties at once from living and nonliving, or, in Vernadsky's phrasing, bio-inert substances. Plant roots share the products of their activity with the soil, in part initiating chemical creations, in part creating fodder for fungi and bacteria. The latter, reworking the remains of plants and animals, taking the sustenance necessary for their life activity, aiding the decomposition of organic material into simpler molecular components which again become accessible for use by plant roots.

Another contribution to breaking down organic material is made by invertebrates—larvae, insects, earthworms, centipedes and millipedes, feeding on fallen leaves and passing them through their bowels. It is estimated that, over a period of 100 years, earthworms pass through their digestive tracts practically the entire soil cover of temperate latitudes to a depth of 0.5 m. In the meantime, they grind down and churn up the soil's mineral and organic elements, improving its structure. The paths they make aid in soil aeration and ease the growth of roots (Lapo 1987; Sorokhin and Ushakov 1991).

Thanks to its ability to accumulate organic and mineral substances as well as moisture, soil serves as a reservoir and source of life for the biota. Soil humus accumulates colossal reserves of carbon and biogenic elements. At the same time, the organic material that accumulates here is different from that contained in plant or animal organisms. Most of all, this is humic acid with its 50–60% carbon content. It is this that gives the fertile chernozem its distinctive black color. Finally, due to its porous, highly dispersed structure, soil has a large surface area of formative particles and, therefore, is able to hold a significant portion of rain and melt water, i.e. to serve as a reservoir of moisture.

Ancient land, we remind you, was deprived of these qualities, so returned all fallen rain unimpeded to the World Ocean in unregulated sheet flows. Thus, in making landfall, the biota oceanized this land by creating soil upon it—a kind of ocean filled with alluvium, as well as the current freshwater hydro system—swamps, rivers and lakes.

***

Let us now turn our attention to the contribution made by the plant biota in regulating the continental water cycle. As we have already said, soil, due to its crumbly structure, represents an efficient mechanism to hold water. But it holds the water only temporarily. Land, after all, rises above the water at a greater or lesser incline. And so, obeying the law of gravity, soil moisture trickles from the more elevated horizon, gathering in streams and rivers small and great before it finally falls into the ocean. The volume of this yearly flow amounts, according to current data, to about 43,000 km^3 (Marfenin 2006, p. 224).

It is estimated that in 4 years, all of continental water accumulated in lakes, glaciers, and swamps would drain into the ocean were it not refilled from atmospheric rainfall. Much of this rainfall forms itself overland, but roughly one-third, or 35 cm of 100 cm of the rain that comes down on land, owes itself to evaporation of the world ocean. In other words, if ocean moisture did not pour as rain or fall as snow upon the land, then in less than 10 years, dry land would justify its name—it would utterly dehydrate.

Earlier on we mentioned the hypothesis that it is forests that enable rainfall over the interior parts of continents, forming "secondary" clouds due to transpiration above dry land rather than ocean. These clouds move with air masses into continents and provide moisture to forests far from the shoreline, unloading rain and snowfall on them as they go. This, in turn, evaporates as accumulated moisture, causing "tertiary" clouds to arise, and so on.

Of course, what Russians call the "weather kitchen"—the mix of atmospheric fronts, storm development and cyclone formation—is an area of very complex and insufficiently researched phenomena, poorly suited to mathematical formulation and modeling. Nonetheless, in terms of the concept of biotic regulation, Viktor Gorshkov and Anastasia Makarieva undertook an attempt, on the basis of known laws of physics, to more precisely describe the transport of ocean moisture to dry land, linking this process to the functioning of the plant biota, i.e. to prove its universal influence on the processes of the continental water cycle. This primarily relates to the preservation of forest lands, and the authors named this mechanism the *forest biotic pump of atmospheric moisture*. But, in order to more fully deal with the substance of this mechanism, we must linger in more detail on the process of *transpiration*, the evaporation of water from the leaf surface of plants, which holds the key to understanding the nature of this phenomenon.

Transpiration, somewhat analogous to blood circulation in animals, is the incessant movement of water together with the organic and inorganic substances dissolved in it up from the soil through the root system of plants and further along the stem vessels of the xylem (the vascular tissue of plants) to the leaves. The vessels of the xylem are tubes with a narrow shaft, whose diameter varies from 0.01 to 0.2 mm. In order to draw water up a large tree through such tubes, it requires pressure on the order of 4000 kPa.[3] But by mere capillary power, even along the thinnest vessel, water cannot rise higher than three meters, while some trees reach 50 or even 100 m in height, such as the Californian sequoia or Australian eucalyptus.

This phenomenon can be explained by the theory of bonding or cohesion. According to the theory, water rises from the roots as a result of evaporation in the leaves, which deprives cells there of water and raises concentrations of dissolved substances. As water leaves xylem vessels in the column of water, it creates tension going down the stem right to the roots. This is connected to the ability of water molecules to bond with one another (cohesion). This property comes from their polarity, a dipole moment, which causes water molecules under the influence of electrostatic forces to attach to one another ("stick," you might say) and stay together in hydrogen bonds. Thus arises the propellant force of transpiration, determined by the falling gradient of hydraulic potential, which falls as concentrations of dissolved salts in the xylary fluid rise. As a result, water from sap with higher hydraulic potential streams to leaf cells, enabled by the selective permeability of cell membranes.

The speed at which water travels along vessels in the stem is notably high. Among grasses, water goes about 1 m/h, and among tall trees—up to 8 m/h. Thanks to cohesion, tension in the xylem vessels carries enough force to pull up the entire weight of the water column. Different estimates of tensile strength for this column of sap vary within the margins of 3000–30,000 kPa. Leaves carry a hydraulic pressure on the order of 4000 kPa. So this column of sap is durable enough, in all likelihood, to sustain the tension it creates (Green et al. 1984, vol. 2, Ch. 14.3.3).

At the final stage, water seeks to abandon the plant, since the hydraulic potential of the surrounding moderately moist air is tens of thousands of kilopascals lower than in the plant itself. The water abandons it in gaseous form, which requires additional energy provided by the unseen warmth of gasification. The sun's rays supply this energy, ultimately serving as the force that moves the transpiration process at every stage—from soil to roots and from roots to stem and leaves.

---

[3] 1 Kilopascal (kPa) = 0.01 atmosphere of pressure.

That plants require water to provide for their life activity, including the needs of photosynthesis, is obvious. Less obvious is the intensiveness of the process. After all, the plant itself holds back only 1% of the water it swallows from the Earth, and the other 99% returns to the atmosphere through the plant, in transit, you might say. Meanwhile, the level of transpiration, with sufficient sunlight, soil moisture and surrounding air temperature, can be very high. Thus, for example, grasses such as cotton or sunflowers are able to expend 1–2 L of water in a 24-hour period, and a 100-year oak—more than 600 L.

Over the course of evolution, the majority of plants have developed adaptations to enable the regulation of this process and retain moisture if needed. Some, for example, cast off their leaves during seasonal chills or droughts. Some put away moisture in secretory cells or in the cell walls of different parts of the plant. Some, finally, developed stomata—special pores in the epidermis, located on the leaves and some parts of green stems, through which gas exchange occurs and 90% of water evaporates. Thanks to special interlocking cells, stomata can close in dry weather or at night when photosynthesis stops, thus slowing the process of transpiration. There are other adaptations for reduced transpiration as well, formed under the conditions of dry climate and water shortages, such as a thickened cuticle (a waxy layer covering the epidermis of leaves and stems), opening stomata at night and closing them during the day, etc. (Green et al. 1984, Vol. 2, Ch. 14.3.8).

Nonetheless, under normal circumstances, due to the great surface area of leaves typical most of all in forest cover, water loss from transpiration can be very high, materially increasing evaporation from the reservoir's surface equal in area to the tree canopy's projection over the soil. And if you consider that total leaf surface for the whole vegetative biota exceeds by four times the area of dry land, it becomes clear that complete evaporation of natural forest in possession of a high leaf index (relative value of exposed leaf area to canopy projection over soil) could successfully compete with open ocean surface of the same temperature. So, according to estimates, maximal water evaporation over forest corresponding to aggregate global flow of solar energy absorbed by the Earth's surface, adds up to ~2 M/year, while evaporation from the ocean surface comes in at nearly half as much: ~1.2 M/year (L'vovitch 1979).

It's also worth remembering that transpiration is not the only source of evaporation, since trees have the ability to accumulate a significant amount of rainfall and snow by catching it in their canopies. This moisture contributes a share of forest-developed evaporation that may reach 30%, which is especially relevant for boreal coniferous forests, where the snowy coats and hats caked on pine trees provide evaporation flow even in winter, when transpiration does not. So virgin forest is capable of evaporating moisture practically year-round, and that from the point of view of the authors of the forest pump concept, carries decisive meaning in the formation of the continental water cycle.

This is because passive geophysical streams, carrying moisture from the ocean, extinguish themselves as they move deeper into a continent. They extinguish themselves exponentially. And the further one goes from shoreline, due to the elevated position of continents, the greater the share of rainfall brought from the ocean that

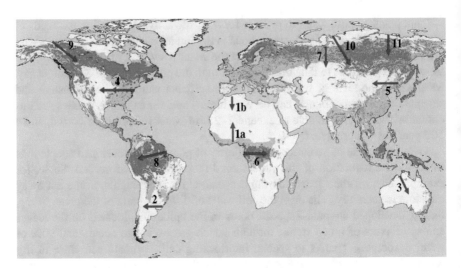

Fig. 11.1 Geophysical regions where research was conducted on the dependence of average yearly rainfall on distance from the ocean. The numbered arrows correspond to the regions: 1a—West Africa, 5° east longitude; 1b—North Africa; 2—South America, 31° south latitude; 3—Northern Australia; 4—North America, 4° North latitude; 5—Northeast China; 6—Africa, Congo Basin; 7—Ob Basin; 8—Amazon Basin; 9—Mackenzie Basin; 10—Yenisei Basin; 11—Lena Basin. Source: Biotic Regulation website: http://www.bioticregulation.ru/common/pdf/06e03s-hessd_mg/06e03s-hessd_mg-screen.pdf

returns to it with the river streams. Granted, this rule holds true, most of all, in deforested areas with low steppe vegetation, where for every 400 km one goes inland on steppe, savannah or prairie abutting the shoreline, the flow of moisture and intensity of rainfall decreases by half.

Data analyzed by Gorshkov and Makarieva on the gradient of decreasing rainfall over wide areas of five continents (Fig. 11.1) has shown that passive geophysical transport of moisture is capable of providing normal conditions for grass canopy and scrub vegetation only adjacent to the ocean to a depth of several 100 km (Fig. 11.2c), mainly during the summer period.

But how do you explain the existence of well-watered areas deep within continents, thousands of kilometers from the ocean—in Siberia, Canada, Alaska, Equatorial Africa or the Amazon Basin (Fig. 11.2e–g)? It would be difficult to answer this question, only going off of passive geophysical streams. Here we must draw our attention to the active transport of ocean moisture, the moving source of which is the forest biotic pump. More exactly, those atmospheric physical processes that occur over forests as a result of transpiration or recapture of rain water by tree canopies.

The essence of these processes is as follows. In a stationary, undisturbed atmosphere at any elevation, air pressure is balanced by an atmospheric air column located above this height. And since an increase in elevation reduces the scale of the atmospheric column above it, the pressure balancing it falls correspondingly. Anyone who has ever climbed a mountain knows this well from personal

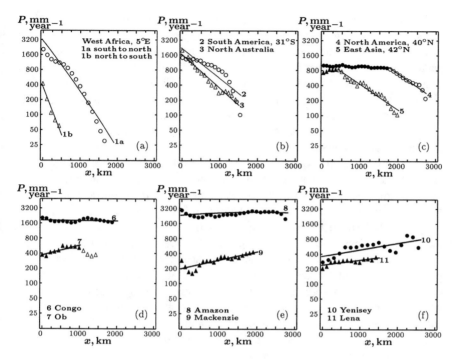

Fig. 11.2 Dependence of the amount of rainfall P (mm/year) on distance X (km) from the ocean. On deforested ((hollow dots/squares) and forest-covered territory (filled dots/squares). Numbered regions are on the map (Fig. 11.1). Source: Biotic Regulation Website: http://www.bioticregulation.ru/common/pdf/06r08o-eopmp_gm.pdf

experience: It's hard to breathe on a mountaintop because the air is thinner. But while other components of air—nitrogen, oxygen, etc.—are only found in a gaseous state, you couldn't say the same of water vapor, which under typical Earth temperature conditions has two phases, liquid (as raindrops and fog) and gas. Because of this, it behaves somewhat differently, i.e. it is able to transition from one phase to the other.

Fog, as you know, forms with lowered temperatures. This phenomenon is called condensation, and we've all had to deal with it before, observing, for example, the accumulation of dew that settles upon the grass on cool summer nights, or some kind of quickly chilling surface, especially a metallic one. This can be explained by the reduction of kinetic energy in water molecules and the slowing of the evaporation process as the temperature goes down. As air cools, the water vapor contained within reaches the saturation point and begins to condense into dewdrops. In the handsome phrasing of Anastasia Makarieva, "Water molecules 'pack themselves' into drops that occupy a thousand times less volume than water vapor—the gas from which the drops form. And since air pressure at earth's surface is proportional to the overall number of gas molecules in an atmospheric column, atmospheric pressure decreases wherever condensation occurs" (From an interview with the newspaper *Nevskoe Vremya*, August 24, 2014).

In physics, the critical temperature at which condensation begins is called the dew point, depending, in turn, on pressure and relative moisture of the air. Something similar happens to water vapor as it rises in elevation, which, as you know, is accompanied by a drop in air temperature—about 6 °C for every kilometer. Thus, for example, at an elevation of 10 km, where modern airliners fly, the outside temperature is nearly 60 °C lower than at ground level. If, like other components of the atmosphere, water vapor were not a condensing gas, its state of equilibrium would continue at any elevation, independent of temperature, and its pressure would decrease by half for every 9 km it rose.

Over the conditions of quickly lowering temperatures, however, in the upper layers of the atmosphere, water vapor reaches the stage of critical saturation just as quickly—roughly double for every 10 °C—much faster than atmospheric pressure falls at these elevations. And since the concentration of gaseous water vapor cannot be greater than saturation, its relative excess immediately condenses, abandoning the gas phase. This, in turn, is accompanied by a decreased weight of water vapor in the atmospheric column, which is already incapable of balancing its predominant pressure in the warmer near-surface layers of the atmosphere, which leads to the occurrence of upward force.[4] It is this force that carries rising currents of moist air, which, lifting to the upper layers of the atmosphere, also condenses, forming clouds and falling as rain or snow (Gorshkov and Makarieva 2007).

And here, you might say, we come to the focal point of the biotic pump concept. After all, if rising currents formed due to water vapor condensation in the upper atmosphere are constantly fed by surface moisture, that means that moist air from neighboring areas where there is less evaporation should be sucked up in its place. And, if, as we have shown earlier, evaporation over virgin forest areas surpasses evaporation over the ocean surface, then, therefore, forests propel ocean moisture deeper and deeper into a continent, compensating river runoff and providing soil with moisture year round. Granted, this would occur only under conditions where forest areas stretch to the coastline, such as occurs, for example, in the basins of the Congo, Amazon or the northern rivers of Russia and Canada, where taiga forests butt up against swampy tundra with ocean access, or, at least, separated from the shore by a distance closer than the exhaustion point of passive geophysical transport (~600 km).[5]

---

[4] This force cannot be compensated by other atmospheric gasses, since, according to Dalton's Law, all of them arrive at equilibrium or are removed from it independent of each other. Therefore, if in some part of the atmosphere a phase change of one of its formative gasses takes place (like water vapor turning to fog), then a rapid fall of atmospheric pressure occurs in that zone. Such as what happens when all the oxygen is pumped out of a vacuum chamber.

[5] There is the particular question of how the great Siberian river basins, thousands of miles wide and covered in forest, can exist by propelling moisture from the frozen Arctic Ocean with its low capacity for evaporation. But the paradox is illusory. It all has to do with the difference in between the intensity of evaporation over the Arctic and the position of the Taiga River Basins in warmer, more southerly latitudes, which ultimately determine the species of horizontal currents of moist air. So the propulsion of ocean moisture is obviously an easier task for the Siberian biota than for tropical forest in the equatorial zone. After all, to provide an analogous transport of moisture from a warm ocean requires significantly higher levels of transpiration.

Thus the destruction of forest cover to a depth of 600 km from the shoreline rips through the biotic pump, and rainfall in the continental interior ceases to compensate river drainage. Moisture in the soil trickles to the ocean, forests dry out and river basins cease to exist. All of these irreversible changes can occur over an extremely short period of time, on the order of 4–5 years—the time required for fresh water, accumulated in mountain glaciers, lakes and swamps, to run out.

In all likelihood, something of the kind took place in Australia 50–100 thousand years ago, when the first humans settled it. We can naturally assume that the new arrivals, as always happens, first took to the shoreline, destroying forests along the way over the whole perimeter of the continent. And when this deforested strip reached the depth at which passive geophysical currents exhaust themselves, the biotic pump was cut off from ocean moisture. Three quarters of native forests then gave way to Australian deserts. By the way, could it be for this reason that most deserts either border on the ocean shore or have access to inland seas? From the position of what has just been said, this detail of geography finds its origin in human history, in human activity, acquiring new territory starting from the seashore.

It might seem that Western Europe, deprived of 9/10 of its natural forests with the exception of northern Scandinavia and the mountain regions of the Alps, Carpathians and Pyrenees but nonetheless free of desertification, would disprove the conclusions drawn above. This, however, is the exception that proves the rule. If Europe has avoided such a fate, it primarily owes this to its unique geographic position—surrounded by internal seas and universal proximity to the shoreline, due to which no territory of this subcontinent is separated from the sea by a distance greater than the point that geophysical transport of moisture is exhausted.

This situation, clearly, gives rise to the illusion that the practice of forest extirpation can be transferred to other regions of the globe with impunity. For these reasons, it will most likely prove, or has already been proven, much more ruinous. By the way, we cannot rule out that the sharp increase in the frequency of catastrophic floods witnessed in Western Europe in recent years might at least in part be linked to the destruction of native forests in mountain regions, which has led to disruption of the natural hydrological regime, the melting of mountain glaciers, etc.

But while we can consider desert practically closed to water (Fig. 11.3a) since the total lack of transpiration there leads not to land sucking up moist air from the ocean but the opposite, dry air being carried out to sea, evaporation may intensify over the ocean's surface in landscape zones of the grassland type, though only in the warm season (Fig. 11.3b, c). During this period, a horizontal current of moist air, commonly known as the summer monsoon (rainy season), arrives from the ocean and gradually wanes over distance. In the colder winter period, evaporation over the grass and scrub becomes less oceanic, and so the remaining moisture is pulled from land to sea, creating the dry winter monsoon. And though vegetation of the steppe type ecosystem provides the support of a certain moisture reserve and evaporation current overland, the lack of forest cover with its high leaf index does not allow this to develop to the level at which moist currents from the ocean would sufficiently compensate river runoff.

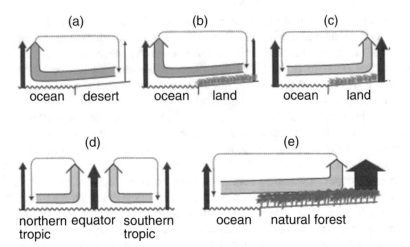

Fig. 11.3 Physical principle of air distribution overland from areas of lowest evaporation to areas of highest evaporation: (**a**) a desert "closed" to water; (**b**) winter monsoon; (**c**) summer monsoon; (**d**) trade winds; (**e**) biotic pump of atmospheric moisture. Thick arrows—horizontal and rising currents of moist air. Thin dotted lines—horizontal and falling currents of dry air. Source: Biotic Regulation Website: http://www.bioticregulation.ru/common/pdf/hess07.pdf

In this way, a fully-functioning biotic pump is the greatest "invention" of the land-borne biota, truly possible only under the conditions of natural primordial forests suited to a given climate zone, whose genetic properties are correlated to the geophysical particulars of the place. Therefore, humanity's primary task, according to Viktor Gorshkov and Anastasia Makarieva, should be recognized as the immediate cessation of the criminal practice of cutting down virgin forest on the territory of river basins, as well as places of access to the shore on oceans and inland seas, with the simultaneous restoration of forest cover on neighboring territories. If this is not the case, we risk not only the loss of the priceless forest wealth we have inherited, but also the conversion of enormous swaths of developed land into barren deserts.

References

Barnola, J. M., Pimienta, P., Raynaud, D., & Korotkevich, Y. S. (1991). CO2 climate relationship as deduced from Vostok ice core: A re-examination based on new measurements and on re-evolution of the air dating. *Tellus B, 43*(2), 83–90.

Bolin, B. (1983). The carbon cycle. In B. Bolin & R. B. Cook (Eds.), *The major biogeochemical cycles and their interactions. SCOPE 21.* New York: Wiley.

Broecker, W. S., Peteet, D. M., & Ring, D. (1985). Does the ocean-atmosphere system have more than one stable mode of operation? *Nature, 315*(6014), 21–26.

Budyko, M. I., Ronov, A. B., & Yanshin, A. L. (1987). *History of the earth's atmosphere.* Berlin: Springer. [in Russian].

Eskov, K. Y. (2000). *The history of the earth and life on it* (p. 352). Moscow: MIROS/MAIK. [in Russian].

Gorshkov, V.G. & Makarieva, A.M. (2006). Biotic pump of atmospheric moisture as driver of the hydrological cycle on land. *Hydrology and Earth System Sciences, 11*, 1013–1033. Retrieved from http://www.bioticregulation.ru/common/pdf/hess07.pdf.

Gorshkov, V. G., Gorshkov, V. V., & Makarieva, A. M. (2000). *Biotic regulation of the environment: Key issues of global change*. Chichester: Springer/Praxis. Also see V. G. Gorshkov's and A. M. Makarieva's site on questions of environmental biotic regulation at http://www.biotic-regulation.ru/pump/pump.php.

Green, N. P. O., Stout, G. W., & Taylor, D. J. (1984). In R. Soper (Ed.), *Biological science* (Vol. 1 and 2). Cambridge: Cambridge University Press.

Koronovsky, N. V. (2003). *General geology* (p. 448). Moscow: MGU. [in Russian].

Krasilov, V. A. (1992). *Natural conservation: Principles, problems, priorities*. Moscow: Institut okhrany prirody i zapovednogo dela. [in Russian].

Lapo, A. V. (1987). *Traces of former biospheres*. Moscow: Znanie. [in Russian].

L'vovitch, M. I. (1979). *World water resources and their future* (p. 415). Washington: American Geological Union.

Marfenin, N. N. (2006). *Sustainable development of humanity*. Moscow: MGU. [in Russian].

Sorokhin, O. G., & Ushakov, C. A. (1991). *Global evolution of the earth*. Moscow: MGU. [in Russian].

Chapter 12
Biotic Mechanisms for Supporting Environmental Stability

From the materials of the previous chapter, you can probably already understand the level of organization the biota imposes on its non-living (abiotic) environment, and the role it plays in forming every component of the biosphere. And one must offer tribute to the remarkable balancing ability of this titanic mechanism in bringing the intricate biochemical and hydrological cycles together into a single whole and providing support for the environmental conditions crucial to life. But while we witness such a harmonious concordance on the greatest scale, it's not hard to guess that it begins at the smallest—from the ecosystem and biocenose, the internally correlated cells of the biosphere. In them, each of the species belongs to a biological community, set within a complex trophic chain through which cycle energy and chemical substances necessary for life.

If we take energy as a base, then this cycle begins, as you know, with plants (producers, photoautotrophs). These are the only organisms, with the exception of a few species of bacterial chemoautotrophs[1], capable of synthesizing complex organic molecules from simple mineral compounds using energy from solar radiation (photosynthesis). It is through them that the flow of energy enters a biotic community, along with the organic materials used by consumer organisms of the first, second and following degrees—herbivores, predators and detritivores—and finally the bacteria and fungi that decompose dead tissue (reducers). Each of the species occupies its own particular ecological niche, which is understood to be not only its physical habitat, but also its role in the community—its feeding habits and interrelationship with other species.

Thus, the roots, trunk and canopy of a given tree grant haven to a great multitude of plant-eating insects and their larvae who eat its leaves, bark and adjacent layers of wood. In turn, these too serve as quarry for predatory insects, birds, and other insectivores. Furthermore, flying insects pollinate the tree as it flowers and birds,

---

[1] These bacteria have the ability to gather energy for organic synthesis from the decomposition of several chemical substances—hydrogen sulfide, ammonia, etc. However, in the overall cycle of matter, they play a relatively minor role.

© Springer International Publishing AG, part of Springer Nature 2018
V. I. Danilov-Danil'yan, I. E. Reyf, *The Biosphere and Civilization: In the Throes of a Global Crisis*, https://doi.org/10.1007/978-3-319-67193-2_12

gulping down its ripe fruit, spread its seeds together with their excrement. When the first, second and third complete their life cycles, the forest "garbage men" get to work. These detritivores–insects and other small invertebrates—feed on carrion or store their eggs within it, providing the newborn larvae their first meal. Finally, at the ultimate stage of this transformation, this remaining undecomposed organic material is reprocessed by the true reducers, fungi and bacteria, who decay it to low molecular-weight compounds accessible to the root systems of plants, which thus restore chemical elements necessary for life to the living from the dead.

Of course, the success and efficiency of this cycle would be impossible unless the scale of consumption for each species was balanced by the consumption of all others. Thus, for example, if the activity of birds and other insectivorous animals and insects were on order lower than that of bark-eaters, aphids and other "wreckers" that often reproduce according to a geometric progression (one aphid, by the end of summer, could produce 13 generations of progeny or 1024 individuals), then an entire forest would be stripped bare from top to bottom in a matter of weeks.

But besides the balance of species within a community, no less important to an ecologist is the interaction between this community and its non-living environment. Here a researcher might allow himself to temporarily forget the existence of distinct species (much as a zoologist, researching the behavior of an animal, does not think about the function of its heart or kidneys) and approaches the biological community as an autonomous functional unit, primarily paying attention not to the particular but to the general features independent of concrete conditions or the geography of the habitat. In ecology this is known as the *ecosystemic approach*, and one of its central tasks is to reveal the fundamental rules that equally govern any ecosystem, even those which differ as much from each other as, let's say, tropical forest, the Eurasian Steppes or the Canadian Arctic tundra.

Let's start at the basics. Like physics, biology has a concept of *work*. Only this is applied not to a machine, but to a living organism. This concept reflects the quantitative characteristics of consumption and reworking of energy in the process of fulfilling one life function or another. In this sense, work can refer to intracellular synthesis, matter transport from one part of the organism to another and transmission of impulses along the nervous system, not to speak of the mechanics of contracting muscles and the body's locomotion through space.

As we've said, this process of transforming energy begins with plants, capable of directly catching the sun's rays, while other living things receive it along with food in the form of chemical bonds with complex organic molecules. At the same time, not only the individual organism but the entire biotic community can be like a mechanism that consumes energy and fodder for the mutual execution of work in the interests of the community as a whole. We could mention its support for the water cycle or the processing of dead organic material into low molecular-weight compounds accessible for use by plants. And as a result, the vector of all these interconnected processes, both of the biocenose (ecosystem) level and for the biota as a whole, is directed at preserving environmental parameters beneficial to life, without which it (life) would be simply impossible.

Supporting conditions beneficial to life, however, primarily means effectively opposing those forces which are ever ready to stamp that life out, or at the very least to expel it from some territory or another. In the language of non-equilibrium systems theory, this effect means one thing: *perturbation*. Perturbation for the biosphere includes sharp cooling of the climate (glaciation), shifts in the concentration of chemical substances necessary for life, hurricanes, forest fires, and so on.

It stands to reason that the biota is incapable of influencing such natural events as volcanic activity, tides or tectonic shifts. However, it can adapt to them, forming corresponding mechanisms able to compensate or tamp down on detrimental consequences of these and other natural events, shifting the balance of nutrient consumption toward neutralizing perturbations as they occur and thus easing a return of the environment to an unperturbed state (analogous to Le Chatelier's principle of thermodynamic equilibrium in physio-chemical systems). And since the biota's basic instrument to affect the environment is the synthesis of organic substances and their destruction, then we might speak of changes in the relative intensity of these two processes in the biosphere (Gorshkov et al. 2000a: p. 110–111).

So, excess carbon dioxide gas in the atmosphere can be absorbed by way of intensifying organic synthesis and transformed into the form of organic carbon. By the same token, a shortage of CO_2 in the atmospheric air could be supplemented by the decomposition of organic stores created earlier and stored in soil humus, peat or organic substances dissolved in the ocean (oceanic humus), where 95% of these substances in the biosphere are concentrated. At the same time, the biota's ability to create raised localized concentrations of nutrients bears unquestionable witness to the fact that synthesis flows and the decomposition of organic material significantly exceeds physical transfer flows of nutrients.

For example, the level to which soil is enriched with the organic and inorganic compounds plants need significantly increases their concentration in lower soil layers where organisms do not live. From this, it follows that nutrient concentrations in the soil are regulated biotically. The same relates to phytoplankton absorbing excess carbon dioxide arising from the ocean depths (a biotic pump). Therefore, and here we observe the same productive role of the biota, the supporting gradient of CO_2 concentrations is an order larger than if it were conditioned only by physical factors—the stirring of the deep water and surface layers of the ocean. In this way, by absorbing carbon dioxide gas dissolved in the ocean, it erects a roadblock to its unencumbered diffusion into the atmosphere, helping to maintain CO_2 concentrations in the air at the level necessary for life.

Another, even more massive reservoir of inorganic carbon, and a source of its entry to the atmosphere, is volcanic activity. Scientists estimate that the power of this geophysical flow is roughly equal to 0.01 gigatons per year. At the same time, global reserves of biospheric carbon make up within an order of magnitude of 10^3 gigatons (Degens et al. 1984; Holmen 1992), and, therefore, this quantity could accumulate through emissions from the inner Earth over the course of about one hundred thousand years. Life on Earth, however, has existed for about four billion years. Thus, over only the Phanerozoic Eon (the last 800 million years), the overall quantity of inorganic carbon in the biosphere should have, theoretically, grown by ten thousand times. As you can see, this did not occur. The reason for this is the depositing of organic carbon in sediments formed in the process of rock erosion. And, as has recently been established, a crucial role in this is played by plants and micro-organisms (Schwartzman and Volk 1989).

As noted above, carbon dioxide, in the process of erosion, dissolving into rain and groundwater to produce carbonic acid, reacts with silicate minerals in rock and is carried out to the World Ocean in the form of bicarbonate ions. Here, after a number of transformations through the sea biota and after it dies away, carbon, now in the form of organic compounds, is removed from circulation and forms seabed sediment deposits. The depth of these deposits reaches dozens of meters in some places, and, as researchers of recent decades have shown, concentrations of these dispersed granules in them are on the order of 10^7 gigatons of carbon (GtC), accumulated over the period of roughly a billion years (Budyko et al. 1987). In this way, the flow of deposited organic carbon in sedimentary rock coincides with the geophysical flow with relative exactitude on the order of 0.01 gigatons per year (Fig. 12.1).

On the other hand, we cannot fail to notice the correlation in order of magnitude between global reserves of organic and inorganic carbon, which speaks of an equality between flows of organic synthesis and organic destruction maintained by the biota to a highly exact degree. Granted, it is not yet possible to measure these reserves directly with sufficient reliability. Thus we can only judge them within the order of magnitude $\sim 10^2$ GtC (Whittaker & Likens 1975; Holmen 1992; Gorshkov et al. 2000b), and by their tendency to change in the past—through indirect evidence. So, for example, research into the CO_2 content in air bubbles from ice cores in Antarctica and Greenland, taken at various depths and, therefore, at different ages, have shown that its atmospheric concentration has stayed more or less constant for the past ten thousand years (Neftel et al. 1982). For times measured in hundreds of thousands of years, atmospheric CO_2 concentrations maintained an order of magnitude (Barnola et al. 1991). Such a correlation, of course, cannot be a coincidence. It bears witness to the enormous potential of the natural biota, providing compensation for environmental perturbations in the interest of maintaining its own stability.

Along with this, on the basis of relative values shown in the diagram, you can calculate the rate at which the biota runs through all the organic and inorganic carbon in the biosphere through the processes of synthesis and decomposition. The relative value for reserves of both one and the other ($\sim 10^3$ GtC) to the global biota's productivity ($\sim 10^2$ GtC/year) characterizes the turnover time for nutrient reserves in the biosphere on the order of less than one hundred years. That is, using only the synthesis of organic substances, all of the inorganic carbon in the biosphere could be expended and converted to organic compounds in the space of mere decades. The inverse also holds true: using only decomposition, all the organic carbon in the biosphere could also be expended in a matter of decades.

Due to this, the question must arise: Why does the biota "need" this enormous and even seemingly excessive biological productivity? After all, to compensate for perturbations such as inorganic carbon emissions from volcanic activity, it would theoretically require a productivity level lower by four orders of magnitude.

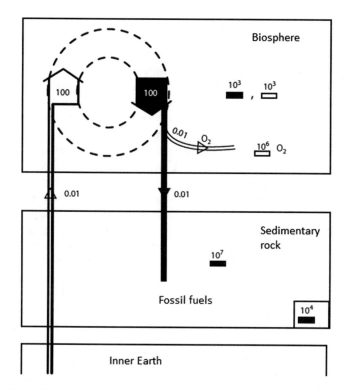

Fig. 12.1 Yearly flows and reserves of carbon in the biosphere, according to Viktor Gorshkov (1995). Carbon reserves are the numbers above small rectangles in units of gigatons of carbon (GtC). Carbon flows are numbers on arrows GtC/year. Flows and reserves of organic carbon are in black or above black rectangles, respectively. Flows and reserves of inorganic carbon are depicted in the white arrow or above empty rectangles. The flow of organic carbon deposits in sedimentary rocks is equal to the difference between its synthesis and decomposition in the biosphere. This flow coincides with a pure flow of inorganic carbon into the biosphere, with relative exactitude on the order of 10^{-4}. The flows of synthesis and decomposition coincide so exactly that they have provided a consistent reserve of organic and inorganic carbon for the whole Phanerozoic Eon (6×10^8 years). At the same time, all oxygen liberated through photosynthesis has accumulated in the environment (underlined by the hollow line and rectangle, number in GtO_2) and is not deposited in sedimentary rock (Gorshkov et al. 2000 p. 117)

However, inconsistency defines geophysical processes on Earth, and, along with more or less regular environmental perturbations, the geological record also contains instances of significantly more serious cataclysms like the great glaciations, sudden outbreaks of volcanic activity or the crash of major asteroids. Therefore, the surplus power of the biota in terms of organic synthesis and decomposition cannot be treated as anything other than an adaptation, kept in reserve, you might say. This allows it, in a relatively short period of time, to also compensate for extreme environmental perturbations and thus provide for the survival of most biological species—as it was, for example, during the last ice age.

The enormous power acquired by the biota, however, presents a certain danger to the environment. When the parity of synthesis and decomposition is violated, the environment can undergo dramatic changes in the space of a few decades. This may occur, for example, in cases of deep change to the internal structure of the biosphere which we will explore later. For now, we will only mention that attempts to artificially reconstitute nature, like the drive to maximize productivity levels in man-made agrocenoses, threaten much greater perturbation and accelerated degradation of the environment than even complete local extirpation of a biota, as in cases of desertification.

The rapid increase in atmospheric carbon dioxide concentrations over the past century stands as indirect witness to this. Until recently this was linked only to combustion of fossil fuels. In answer to such an environmental perturbation, it would seem the biota, reacting in accordance with Le Chatelier's Principle, should swallow up the excess carbon dioxide gas accumulated in the atmosphere. A global analysis of land usage shows, however, that on land under human cultivation, the quantity of organic carbon accumulated by disrupted ecosystems is not increasing but decreasing (Houghton et al. 1983, 1987). Meanwhile, the rate of carbon emissions into the atmosphere from the perturbed continental biota corresponds within an order of magnitude to the rate of fossil carbon emissions from combusting coal, oil and gas (Watts 1982; Rotty 1983). How this threatens the biosphere and what practical conclusions we should draw from this will be discussed in detail in Chap. 15.

***

But how can the biota maintain so exactly the parity of synthesis and decomposition of organic matter over the course of millennia and even geological epochs? After all, the unusual complexity of life expressed at the biomolecular, cellular and organism level is at once its Achilles heel. For the more complex the organization of a given system, the more vulnerable it is to gradually mounting disorder (entropy) and the more inevitable its chances of degradation and collapse. This rule proves true even for the genetic program of an organism, that guarantee of reproducing life over generations, which is also vulnerable to mounting destructive changes appearing among the progeny of each individual. At the same time, the number of defective individuals could be viewed as a specific characteristic of a species. Thus among humans, for example, one of 700 newborns suffers from Down's Syndrome, and of a hundred people living to age 55, one will come down with Schizophrenia, the predisposition for which, as you know, is genetic.

Among the above-mentioned types and levels of correlated life systems, we do not see great distinctions in biocenoses, the elementary cells of the biosphere where each species occupies its own ecological niche without overlap from other species and fulfills its own specific share of work to stabilize the environment. Correlations within a single given community, however, can be rather strict. Each of the ten-thousand-plus species of lichen, for example, is not an organism at all, but a symbiosis of organisms: a very particular sort seaweed and a particular type of fungus (Farrar 1976). Some insects can feed on only single, lone species of plants (Raven and Johnson 1998), while the flowers of some of those plants can be pollinated only by a particular species of butterfly, etc. So in a normal, unperturbed community, there is no interspecies competition, and, thanks to almost total impermeability of the cycle of matter, practically no waste. Specifically, it is the need to maintain a high degree of isolation in the matter cycle that dictates the need for the existence of sustainable biological communities.

But while the correlation of a biogeocenose community is brought about by the necessity of maintaining parity between organic synthesis and decomposition, that very correlation also serves as the cause of its relative short-livedness and inevitable collapse with the passing of time. Collapse of a community owes itself to the accumulation of mutant individuals diverging further and further from the species standard and the gradual weakening of the correlation of species, which, in the struggle for food resources, begin to occupy overlapping ecological niches. Such a community, being already in no condition to maintain the stability of the local environment, loses its ability to compete and ultimately disappears from the face of the Earth.

This contradiction between the finiteness of both individual and "collective" organisms (biocenose) and the infiniteness of life in the whole of nature is resolved along a path both prodigal and the only possible: on the basis of competition and selection of autonomous individuals (within a single species) or independent biological communities (within an ecosystem). Thus, among the population are preserved only individuals with an undistorted genetic program and only communities with a species structure that preserves the ability to successfully compensate for chance fluctuations and perturbations in the environment.

Interestingly, humanity came to this very principle of competitive interaction as a result of its many-thousand-year social evolution. This is how the free market functions, crowding out inefficient producers. Extending this principle to such natural structures as biogeocenoses and ecosystems, Gorshkov came to a conception of its universality for nature and society. And it's no coincidence that he named one chapter of the monograph, *Physical and Biological Bases of Life Sustainability* (1995), "The Biosphere as Free Market."

If the reader has not forgotten the barrage of criticism that evolutionary biologists once unleashed on the "Gaia Hypothesis," then now would be a good time to recall the main points emphasizing a distinction between Lovelock's concepts and Gorshkov's. Lovelock imagined a grandiose global mechanism uniting living and nonliving components into an indivisible whole in the interests of preserving the planetary environmental parameters necessary for life. Gorshkov conceives of a "biotechnology marketplace" formed of a great multitude of biological "players"— communities and individuals. Lovelock—the colossal complexity of energy and matter flows penetrating the biosphere, maintained over the course of tens or hundreds of millions of years, but at the same time initially unsustainable and doomed to inevitable collapse and death. Gorshkov—the necessity of permanently maintaining "ecological solvency" and of the right to a place in the sun for each separate individual and each local community as they are run through the sieve of competitive selection, thus attaching to progeny their own species and genetic makeup. In this way, according to Gorshkov, nature "imposes order," working upon an uncounted multitude of independent operating units on the basis of the statistical law of averages, thus minimizing chance fluctuations that threaten the existence of any complexly organized system.

And if we descend one or two steps lower, from the biological community to the separate individual, then we can give examples of how the biota resolves problems analogously at the organism level. For example, the "distributive" circulatory system in animals, using many mutually uncorrelated blood vessels with the aim of dependably providing cells with oxygen (erythrocytes) and disarming elements alien to the body (leukocytes). Or the chaotic mass of randomly oriented leaves on trees and bushes, allowing for maximal catching of the sun's solar rays, and so on.

The idea of the biota's sensitivity to environmental perturbation occupies a special place in the concept of biotic regulation so we will look into it at some depth. The thing is, like the "Gaia hypothesis," Gorshkov's theory also has its thorny spots from the evolutionary biologist's point of view, though of a different order. As we have noted, classical evolutionary theory focuses its attention upon the fate of the individual, the isolated exemplar jokingly dubbed an individualist. But, as we say in Russia, in every joke there is a bit of truth, and this is no exception. Natural selection in the Darwinian understanding has to do with manifold variations within a single population (the species) and the differing degrees of success in adapting to the changing conditions of the environment which provide a greater chance of survival (and preservation of a genotype in offspring) to some and deny such chances to others. At the same time, that which is adequate for a changing habitat has nothing to do with what kind of changes occur, or if they ultimately threaten the well-being of the population or the community as a whole. It is enough for them to survive for the moment and adapt to the concrete conditions that arise.

Things stand entirely differently when the criteria for selection is the capacity to perform work to stabilize the environment. It would seem that selection should make no distinction between communities or individuals forming them, dependent on how well or poorly they execute this mission of theirs. After all, if environmental conditions get worse, both one and the other would find themselves in the same disadvantageous position. And, furthermore, communities that successfully work "for the common good" are also using a portion of their energy resources, and should lose out in competition with those communities that economize them. How then can they survive in the struggle for a place in the sun, and why, in the billions of years that life has existed, has its capability for biotic regulation not disappeared in the endless chain of mutations passed from generation to generation?

To find a way out of this theoretical dead end, Gorshkov proposed adding the concept of *biotic sensitivity* $-\varepsilon$ to external perturbation. According to this idea, the biota reacts only to those changes in the environment that surpass a certain particular ε point (understanding that as the level at which an environmental parameter diverges from its average value).

Probably some of our attentive readers, having visited the forest in summer, managed to notice that breathing feels different under the dense cover than it does in an open, freely circulating field. Even temperature and moisture there distinguishes itself from that of neighboring tracts. Within their canopy cover, trees are able to support their own microclimate and maintain soil conditions on the area of their root systems (Gorshkov and Makarieva 2007). Soil scientists have cast their gaze elsewhere: when moving from tree to tree, cross-sections of soil even from a single source rock have boundaries clear to the naked eye in color, structure and texture. But as we've already established, soil forms through the collective work of all components of a biological community, including bacteria and fungi. If the results of this work can vary even between the territories of neighboring trees, that confirms the fact that each such community acts as an autonomous cell of the biological cycle.

Of course, a local environment on the scale of a single tree and its correlated soil biota is incomparable with the internal environment of an animal's body maintained in a homeostatic regime. The former is blown by winds and washed by storms, and so various fluctuations are practically inevitable. And furthermore, the processes of nutrients, dissolving in the atmosphere and physically mixing with the soil, at first glance seem to nullify the possible distinction of a local "microcosm" of biocenosis communities. And yet the ability of separate mature trees to form an internal atmosphere of a canopy in which the carbon dioxide gas content, for example, can differ from average atmospheric concentrations, still goes beyond doubt. And the whole problem is in how the biota reacts to such differences.

According to the assessments, the biota's sensitivity to changes in the majority of parameters corresponds to a magnitude of 10^{-2} to 10^{-3} (Gorshkov et al. 2000a: pp. 70–71; Gorshkov et al. 2004). Thus, if a shift in the concentration of carbon dioxide gas detrimental to the community in atmospheric air adds up to less than 10^{-2} –let's say 1/1000 of a percent—then the biota will not notice the change and not react to it. At the same time, a difference in CO_2 concentrations of one percent may be critical for a normal community and lead to its functional restructuring. This restructuring could express itself, for example, by depositing excess carbon in organic soil humus, or, in the case of a CO_2 deficit in atmospheric air, to intensifying processes of destruction and release of inorganic carbon. The same relates to soil quality maintenance and the surrounding air, on which a single tree can exert its influence to a certain extent, either to increase or to decrease. It does this by changes to transpiration, the vertical temperature gradient under the canopy, emission of nutrient aerosols into the atmosphere and other, still little researched, processes (Gorshkov and Makarieva 2007).

In this way, a community, having a sensitivity of $\varepsilon \approx 10^{-2}$, acquires a small but noticeable advantage when compared to mutants with a sensitivity point of $\varepsilon > 10^{-2}$, unable to maintain the local environment at settings beneficial to themselves. This ability of trees to maintain sensitivity at a level of $\varepsilon \approx 10^{-2}$ affirms itself genetically in the process of individual selection. As a result, mutant trees, having lost this ability and now inadequately sensitive ($\varepsilon > 10^{-2}$) are gradually pushed from the ecosystem.

Now let us imagine a situation when, as a result of some destabilizing effect, for example, volcanic emissions of carbon dioxide gas, its concentration in the atmosphere materially surpasses the optimal mark for the biota. In that case, with the corresponding level of sensitivity in the biotic communities that make up an ecosystem, this inorganic carbon begins to get absorbed and is converted to an inactive organic form. And if the overall area occupied by such an ecosystem is large enough, a globally significant physical flow of the nutrient comes about from the external environment into the arena of life functions. Obviously, this flow will exist until CO_2 concentrations within and without the ecosystem match each other with the exactness corresponding to the biota's sensitivity, that is, until the global nutrient concentration in the external environment reaches a value beneficial to the environment. Such is the basic scheme of the biotic regulation mechanism, passed from generation to generation in the process of competition and selection of specific biotic communities.

***

One argument in favor of biotic regulation of the environment comes in the form of Henry Cowles' discovery of and Frederic Clements' further research into the phenomenon of ecological succession. Succession is the process of an ecosystem's evolution, distinguished by sharply delineated stages and the seemingly pre-programmed replacement of one group of dominant species by another.

So, for example, a newly formed volcanic island is first colonized by blue-green algae (cyanobacteria) and pioneer communities of lichen, which have no need of soil cover. The space of several decades will pass before they form a layer of soil on which more complex organisms can find suitable conditions. At first this could be moss or non-vascular plants, followed by grasses, then still later by bushes, and, finally, trees. And each previous community seemingly leads the following by the hand, surrenders its place and passes along the baton. At the ultimate stage of succession, a sustainable and self-sufficient community forms, which, barring external disruption, is capable of sustaining equilibrium with the environment indefinitely so long as biomass and population density of specifically developed species remains constant. Examples of communities having completed succession, known as climax communities, include oak forests in wet clay soils or pine and fir forests typical of the European north growing in sandy clay and loam.

The successive settlement of a bare volcanic surface introduced above could serve as an illustration of what is called primary succession. But analogous gradations can be observed in the process of secondary succession, during the restoration of a forest after logging or fire, for example.

So, for 30 years after a forest fire, one can observe on its location total chaos in vegetative cover and maximal entropy in the distribution of productivity to various species of shrubs and trees. In this period, trees grow at top speed, and their ability to regulate the local environment is temporarily at a minimum. Such fast-growing forests, having not yet accumulated the dead organic material through whose decomposition the return of carbon dioxide gas to the atmosphere occurs, are particularly active in depositing carbon, which is very important from a global perspective. At the same time, due to a lack of old, dying trees, the permeation of the matter cycle in such communities can reach levels in the tens of percentage points (Gorshkov 1980).

Only with the passage of several decades after a perturbation does this disruption, as shown by measures of productivity, biomass growth and changes in inorganic substance concentrations in the soil, come down to a few percentage points (Bormann and Likens 1979). And after another 50–70 years, the productivity of the damaged community restores itself, along with its leaf cover and overall nutrient cycle with maximal concentrations in the upper soil horizons. Finally, after a space of 150 years since the damage occurred, the majority of its community characteristics have restored themselves—its biomass, the thickness of soil debris cover, the content and distribution of chemical substances, and also the impermeability of its nutrient cycle. The ultimate restoration of a forest is marked by the formation of a tree layer of uneven age structure, which occurs only 2–300 years after logging of fire (Finegan 1984).

The chemical makeup of the environment, as well, undergoes material changes in the course of secondary succession. This primarily concerns the soil. Local concentrations of various biogenic elements in the soil may change tens or hundreds of times, conditioned by the life activity of species determining the direction of successive changes. Such species, considering their role in the rebirth of the ecosystem, Gorshkov proposed calling reconstructive.

Among boreal conifer forests, for example, reconstructive species would include the birch, the alder, the aspen, berry plants, mushrooms, and many of the animals that feed on these species. The most notable particularity of reconstructive species is their ability to shift concentrations of environmental food sources in a direction that is disadvantageous to themselves, but beneficial to the incoming generation. It is this that explains the phenomenon of graduated succession—the removal of the presently reigning reconstructive community and the arrival of the next reconstructive generation once it has found optimal conditions for itself, in order to surrender its place to a new dominant group in due time. Finally, at the last, pinnacle stage of succession, the concentration of nutrients in the local environment reaches a value advantageous to the climax species and relatively disadvantageous to reconstructors. In this way, the destroyed community returns to its starting point—the sustainable climax state. Here are some of its features:

- Accumulation towards the end of secondary succession of greater, greater and greater share of available food supplies as community biomass and simultaneous depletion of abiotic system components—water and soil mineral plasts.
- An increased quantity of detritus production.
- Detritus turns into the main source of food supplies in the ecosystem, and detritivores—the main consumers, in place of herbivores (Green et al. 1984, vol. 2, Ch. 12.5.2).

Under these circumstances, climax species acquire maximal competitive advantage, establishing a sustainable population that is capable of maintaining this vigorous regime for an indefinitely long period of time. As regards the reconstructive species, they also remain in the climax community, but only in the form of isolated "marginal" individuals. They make up a decidedly sparse population under the restrictive weight of an environment ill-suited to them. And so it remains until the next cycle.

Such, in general terms, is the process of succession strictly specific to each climax community but unfolding according to the single described scheme, independent of geographical location. However, all this holds true only in the absence of regular perturbations, which can not only put the brakes on secondary succession but cut it off all together. If the perturbations take on a systematic character, then it will cause irreversible harm to the ecosystem, which will forget to program for its restoration and never again return to the climax phase. We observe this, for example, during regular forest logging for industrial lumber or when it is systematically worked with herbicide to exterminate low-value types of trees, as well as artificially cutting back or clearing mature forests of over-mature trees and fallen or rotting trunks.

This last factor, by the way, is the most flagrant and dangerous interference in the life of a natural community, since it is the mature forest that represents the healthiest body of the biosphere, in which, when the matter cycle is completely balanced, there is not and cannot be anything "extra." Foliage growth is limited by fungi and bacteria and all organic components of both strong and over-mature trees go toward the process of life activity of other organisms. In this way, the widespread practice of periodically cutting at a typical interval of 50 years literally severs the process of restoration in primordial climax forests with their closed matter cycle and ability to compensate for environmental perturbations. Therefore, for a return to an unperturbed state of the biosphere, the interval between successive clear cutting of forests should be increased to at least 300 years, that is, slowed by six times. And considering that clearing today usually surpasses the volume of natural growth, we ought to be speaking of a reduction in logging on a global scale by a minimum of eight to ten times. (For more on that, see Chap. 15.)

***

There's no need to remind you that the stages of succession described above could not repeat themselves over the course of millennia were they not fixed in the genetic memory of the biota, and that means in the genome of each individual species. So, for example, all reconstructive species within a given succession are programmed to change the environmental parameters toward a direction disadvantageous to themselves and advantageous to climax species. Though, considering the particular role of the latter in maintaining environmental stability, it's not hard to understand that not only the biota as a whole, but the reconstructive species themselves win on this in the end. Accordingly, the ability of the climax community to maintain beneficial conditions for all living things is inseparably connected to the corresponding genetic informatics and a specific selection of biological species in whose genetic memory it is written.

Genetic memory, however, just like any other ordered information, is vulnerable in time to gradual destruction and collapse. Therefore, when speaking of the biota's ability to maintain the preferred environmental conditions, we cannot ignore the mechanism that enables the preservation of this genetic program through the process of its inheritance. According to the concept of genetic regulation, this mechanism, as stated above, is the competition and selection of individuals and their communities.

Evolutionary theory, as you know, designates several types of natural selection depending on the tasks that changing environmental conditions put before a population—directional, stabilizing, disruptive, etc. The concept of biotic regulation primarily addresses stabilizing selection, aimed at conserving average phenotypical markers and thus providing populations the fitness for their usual survival conditions. Filtering out individuals with extreme divergences in phenotype, it blocks the removal of genetic information due to chance mutations at a population level, maintaining the order of the system and preventing the build-up of entropy.

But the selection of individuals is to a certain extent a measure of their quality, that is, their fitness to perform some kind of biological work or another. And, like any process of measurement, it should obviously have some capacity to judge reacting, for example, to mutations of the genome beyond a certain benchmark. Individuals with clearly altered genetic programs and expressed anomalies leading to a decreased competitive advantage are squeezed out of the population, while others, whose changes are below this benchmark, successfully pass through the sieve of stabilizing selection, clearing the ground for intraspecies genetic variation.

The existence of this benchmark of reaction to mutations allows us to explain the phenomenon of *discrete species*. After all, if the course of evolution is uninterrupted in time, and species constantly adapt to changing environmental conditions, then what causes the absence among them of intermediary or transitory forms observed in both modern material and paleontological data? But it all falls into place when interpreted in light of the stabilizing selection described above, which doesn't "notice" immaterial divergence in phenotype, but hems away any that goes beyond a specific species benchmark. At the same time, the existence of this benchmark gives us the key to understanding the surprising persistence of species, comparable in longevity to geological epochs.

There are still possible situations when stabilizing selection seemingly retreats to the background, allowing space for other forms of natural selection to come to the foreground of life. Such occurs, for example, when the regulatory capacities of the biota are depleted at critical stages of its historical development. As we have already noted, a large number of abiotic processes exist both within the Earth and in space that are beyond the scope of the biota's regulatory influence. One of the clearest examples is the transformation, occurring two billion years ago, of the Earth's reducing atmosphere to an oxidizing one, when the biosphere, in the words of microbiologist Grigory Zavarzin, "turned itself inside out," changing fundamentally to a high-nitrogen oxidizing atmosphere with a few oxygen-free pockets where anaerobic micro-organisms found refuge (Zavarzin 2001).

The cause of this was the formation process of the Earth's core, where, by force of gravity, the majority of the planet's iron displaced itself, consequently reducing sharply concentrations in seawater of iron oxide (FeO). And, while in the previous 1.5–2 billion years, all of the oxygen formed by the life activities of anaerobic prokaryotes had been expended on oxidizing atmospheric gases (NH_3, CH_4, CO, H_2S) and iron oxide diluted into seawater, then the liberated oxygen began to accumulate in the atmosphere, which told upon nearly the whole prokaryotic biota, which in its masses was unadapted to life in an oxidized environment. As a result of these cataclysmic events, a global transition occurred in the species makeup of the Earth's biota, and the place of the previously dominant anaerobic microorganisms was taken by the at that time relatively scarce photosynthesizing cyanobacteria (blue-green algae, the most ancient of prokaryotes), which used water and carbon dioxide gas to construct organic molecules, and for an energy source—the visible part of the solar spectrum.

But even incomparably smaller-scale transformations of the environment, accompanied by mass-extinctions of species, have not occurred too often over the course of Earth's history—on average, once in a hundred million years over the last half-billion year period (Raven and Johnson 1998; Jablonsky 1994). Meanwhile, the time required for a transition in the biota's species makeup is calculated in millions of years and takes up whole geological periods.

Unfortunately, the process of environmental degradation and accompanying loss of biodiversity as a result of human economic activity that we observe today has already become comparable to the rate of biota transformation in past geological epochs. But the time frame is incomparable, distinctive by several orders of magnitude. How on Earth can the biota respond to this? Perhaps through the development of new species, which, according to paleontological data, requires tens of thousands of years of evolution? Obviously not, though the theoretical possibility of new species development, especially among bacteria, in response to anthropogenic change to the environment cannot be ruled out. Far more real today is the threat of genetic disorganization in existing species and consequent loss of genetic memory of the biotic regulation mechanism passed from generation to generation.

This is because stabilizing selection is truly effective only under conditions of a natural ecological niche for each species. Individuals with a normative or insignificantly changed genetic program possess the greatest competitive advantage and form a population whose genetic memory saves information of species' properties corresponding to the interests of environmental preservation, as well as of the environment itself and its provision of the species' survival needs.

As natural habitat conditions disappear, however, and genetically programmed methods of responding to external pressures become inadequate to the new reality, such individuals quickly lose their competitive advantage, giving them a green light to disrupt the genome and change genetic memory. This relates not only to domesticated animals or cultivated plants, already long torn from their natural roots, but also to a multitude of *synanthropes*, species closely linked to humans whose ecological niche has been deformed by conditions civilization has brought about. Such, for example, is the house mouse, now incapable of returning to its natural state, or sparrows, having increased their numbers by several orders of magnitude and also almost never encountered outside the zone of human habitation.

We see a clear analogy in forests intensively exploited by humans, which are already practically incapable of returning to climax phase since genetic information of the optimal environment for climax species has been irretrievably lost. And as humans artificially maintain reconstructive species they find pleasing, the community truly loses its capacity for biotic regulation. Should humans "conquer" the whole biosphere, this mechanism could be lost on a global scale. Then, clearly, nothing will be left to us, we "lords of the planet," but to take environmental management into our own hands. That is, to replace biotic regulation with technological. But how much does this correspond to our real capabilities? This is how we must interpret the question of biotic regulation of the environment.

***

Life, as you know, is a process characterized not only by the acquisition and reworking of matter and energy, but of information. In both the rate of information flow and the efficiency with which it is reworked, there exists between the biota and civilization an impassible abyss. Thus, for example, the flow of information (matter exchange) in a given bacterial cell (10^8 bits/s) could be compared with the information flow of a personal computer. Let the molecular "memory units" play the role of logical operators, and the cell itself serve as control panel. For each square micrometer of the Earth's surface, there are several living functioning cells—Plankton in the ocean, plants, bacteria and fungi on land—that non-randomly react to local changes in the environment. The overall quantity of bacteria on Earth is estimated at 3×10^{27}, and the number of cells in the biosphere is roughly on order of magnitude larger. Thus, the flow of information processed by the Earth's biota adds up to $10^8 \times 10^{28} = 10^{36}$ bits/s. This process of data conversion takes place with nearly no energy usage, i.e. with an energy conversion efficiency close to 100% (Gorshkov et al. 2000b: pp. 211–212).

Modern computers, whose aggregate storage space allows us to preserve all of humanity's cultural information, are marked by their speed and high energy conversion efficiency. Nonetheless, next to molecular technologies, their capabilities stand abysmally low. If you gave a computer performing 10^{11} operations per second to every person on earth, the total flow of processed information would not exceed 10^{20}–10^{21} operations per second, which is 15 orders of magnitude lower than in the biosphere. As regards energy efficiency, the most powerful modern computer, capable of performing 10^{16} operations per second, uses about 10^7 watts, and energy usage comes to $10 \times {}^9$ J—12 orders more than in the biosphere. If you covered the whole Earth in supercomputers, each of them occupying an area of 100 m², the total information processing flow would add up to 5×10^{28} bits/s—20 million times less than in the biosphere. And the energy used by such a computer network would go a hundred thousand times beyond that used by the biosphere (Makarieva et al. 2014).

In all likelihood, given the current rate of technological progress, the gap between information flows in the biota and civilization could be reduced by five to six orders of magnitude in the foreseeable future as computers grow faster and more numerous. But even if we managed to close the gap entirely, it would still not solve the problems or allow us to create a technological management system for the environment equivalent to biotic regulation. In part, this is because interaction between next-generation computers and the environment would be qualitatively different from what happens in a living cell, where molecular memory units are integrated into their environment. And this holds true not only for unicellular organisms, but for fungi and higher plants that sustain this quality due to their highly efficient surfaces—spindly, branching fungal hyphae, high leaf indexes, extensive root systems, etc.

But, that's not even the most important part. What's most important is the limited potential of the human brain, particularly sharply illustrated by our interaction with computers. To demonstrate this thesis, let us recall the well-known problem of automatic and manual control.

Manual control takes place on the basis of inborn and acquired information, as well as peripheral impulses coming in along feedback channels from the sensory organs, and is limited by the information processing speed of the central nervous system. Automatic control, based on computer programs, takes place at a speed a million times surpassing human potential. At the same time, the latter must be absolutely sure of correct input in the computer program, testing it many times in the course of preliminary experiments. And, nonetheless, various unforeseen situations often force a person to take control into their own hands, leaning on personal experience, knowledge and intuition, even at the expense of speed to the operation.

From this perspective, you could view the biosphere as a globally distributed system of microscopic computers, with biotic regulation equivalent to a control panel, in which the rate of information processing surpasses human mental capabilities by 30 orders and change, and by ten to fifteen orders—computerized control capabilities. In essence, it serves as the environment's automatic control system, based on programs developed over the course of several billion years. Paleontological data bears witness that roughly once in a hundred million years, a transition of the Earth's biota occurs, accompanied by a mass extinction of old species. Gorshkov supposes geophysical and extra-planetary factors created conditions for these changes. That means that over the past billion years, environmental control programs have been tested no more than ten times. Each program was unique in its epoch, supported by the biota for the longest possible period of time. New biotic programs underwent, through the process of evolution, an experimental trial of many thousands of years, at once preserving the continuity of life's universal biological organization.

Humanity, therefore, according to Gorshkov, in seeking an adequate replacement for biotic regulation of the environment, would need tens if not hundreds of thousands of years, since testing and correction of such programs necessarily comes into being under manual administration. But people do not have the kind of time on their hands that they would need to create a technological control system for the environment. The process of anthropogenic degradation of the biosphere is unfolding far faster, counting down years in the hundreds.

And people shouldn't be setting such goals for themselves anyway. Just the opposite, doing justice to the biota's great perfection, we ought to do everything in our means to preserve it and restore as much as possible of what we have destroyed in our millennia-long barbarity against nature. Then we wouldn't need a technological medium for environmental regulation at all. And we could find more reasonable uses for our growing power.

It would be hard to find a serious ecologist unwilling to subscribe to these words. And nonetheless, in finishing this section which illuminates the key ideas of the concept of biotic regulation,[2] we would err against truth if we limited ourselves to only one side of the coin. Because not all biologists and evolution specialists,

---

[2] We will say more on the biotic regulation concept's handling of the biosphere's carrying capacity in Chap. 14.

unfortunately, share this view. Many look upon it with circumspection, seeing a certain tendency for oversimplification.

Biologist Nikolay Marfenin, in a letter to one of this book's authors, wrote, "The theory tacitly implies that after the biotic processes, the abiotic are all clear and accounted for. But no, it is the abiotic processes that are the greater quandary, still researched very inadequately and so not accounted for. You can't make conclusions about the role of the biota from calculations of the carbon cycle, because the role of the *abiota* remains insufficiently clear." Famed microbiologist and member of the Russian Academy Grigoriy Zavarzin, in his article, "The Antimarket in Nature," while in many ways showing solidarity with Gorshkov ("The description of the community as a holistic evolutionary unit closely coincides with my own understanding of macroevolution's central issue"), nonetheless characterizes his approach as "an attempt to translate the processes of evolutionary biology into the language of university physicists (Zavarzin 2007).

Academy member Nikita Moiseyev addresses nearly the same point in his article, (from "Ekologia i zhizn'," 1998, No. 2), where he characterizes Gorshkov as "a remarkable researcher, having developed a grandiose theory of 'biotic regulation' parameters for the biosphere within whose bounds (quite broad, by the way) it is necessary to support life. But, as often happens with leading scientists, his own scholarly interests fill up the horizon, leaving out many important circumstances in the biosphere's development...And if we look (at it) from the overall systems point of view that we need to, inherent to the process of self-development of such a complex non-linear dynamic system, which the biosphere is, then we see a picture that doesn't look much like the one drawn only through the use of biotic regulation theory."

The majority of evolutionary biologists also do not share the view of the hypertrophic role it assigns to stabilizing selection at the expense of other evolutionary mechanisms (see, for example, (Lima-de-Faria 1988; Chaykovsky 2010; Markov 2015)). On the other hand, it hardly satisfies to explain evolution by way of influence only from external factors on the biota, whether extra-planetary or geophysical. And if you start from the proposition that biotic community functions as a whole submit to the interests of maintaining "determined" conditions of life on Earth, then how do you explain, for example, the origin of the unbelievable variety of species, or such phenomena as preadaptation?[3]

And yet it is for good reason that we have assigned such a substantial portion of our book to this concept. It comes down to the fact that there are not many theories in our day that we might call so essential as the concept of biotic regulation of the environment or the "Gaia Hypothesis." Both one and the other contain no shortage of productive ideas, and even if they are not the ultimate truth, they nonetheless bring us materially closer to it, or at the very least allow us to come closer. Beyond that, each of them presents a fresh, substantive look at the processes of transforming

---

[3] An evolutionary paradox linked to the functional reconstruction of organs that, at the time of appearance, do not have the adaptive value that they receive in the course of further evolution. For example, the swim bladder in fish reconstituted itself as the lungs of land animals.

matter and energy in the surrounding natural world, providing us plenty to ruminate upon. This relates in part to the idea of sustainable development, which within the framework of the biotic regulation concept receives new reinforcement, especially with regard to the preservation of natural ecosystems and forests in particular (more on that in Chap. 15). Perhaps for the first time in the history of scientific thought, the role of inviolate ecosystems is being assigned the pride of place that it rightfully deserves.

Only one thing stirs a reflexive sense of perplexity. However you may relate to Viktor Gorshkov's theory, hiding it under a bushel is unacceptable under any circumstances. Truth, as you know, is born of argument, but no serious discussion has of yet touched upon this theory, though 20 years have passed from the moment of its publication. Such a state of affairs could hardly be called rational. And so we'd like to think that this book will make a contribution to overcoming the incomprehensible "conspiracy of silence." The more this work becomes known, not only to specialists but to everyone concerned with the worrying state of the environment, the better. The wealth of ideas laid forth within it provokes serious consideration forcing us to look upon the delicate natural world that surrounds us in a new way and to recognize the fateful role that humanity's prodigal attitude may play in its fate.

References

Barnola, J. M., Pimienta, P., Raynaud, D., & Korotkevich, Y. S. (1991). CO_2 climate relationship as deduced from Vostok ice core: a re-examination based on new measurements and on re-evolution of the air dating. *Tellus, 43B*(2), 83–90.

Bormann, F. H., & Likens, G. E. (1979). *Pattern and process in a forested ecosystems*. New York: Springer.

Budyko, M. I., Ronov, A. B., & Yanshin, A. L. (1987). *History of the earth's atmosphere*. Berlin: Springer.

Chaykovsky, Y. V. (2010). *Zigzags of evolution. Development of life and immunity*. Moscow: Nauka i zhizn. (in Russian).

Degens, E. T., Kempe, S., & Spitzy, A. (1984). Carbon dioxide: A biogeochemical portrait. In O. Hutziger (Ed.), *The handbook of environmental chemistry* (Vol. 1, pp. 125–215). Berlin: Springer.

Farrar, J. F. (1976). The Lichen as an ecosystem: Observation and experiment. In D. H. Brown, D. L. Hawksworth, & R. H. Bayley (Eds.), *Lichenology: Progress and problems* (pp. 385–406). New York: Academic Press.

Finegan, B. (1984). Forest succession. *Nature, 312*, 103–114.

Gorshkov, V. G. (1980). The structure of biospheric energy flux. *Botanichesky zhurnal, 65*(11), 1579–1590 (in Russian).

Gorshkov, V. G. (1995). *Physical and biological bases for sustainable life*. Moscow: VINITI. (in Russian).

Gorshkov, V. G., & Makarieva, A. M. (2007). Biotic pump of atmospheric moisture as driver of the hydrological cycle on land. *Hydrology and Earth System Sciences, 11*, 1013–1033. Retrieved from http://www.bioticregulation.ru/common/pdf/hess07.pdf.

Gorshkov, V. G., Gorshkov, V. V., & Makarieva, A. M. (2000a). *Biotic regulation of the environment: Key issues of global change*. Chichester: Springer/Praxis.

Gorshkov, V. G., Gorshkov, V. V., & Makarieva, A. M. (2000b). *Biotic regulation of the environment: Key issue of global change*. London: Springer.

Gorshkov, V. G., Makarieva, A. M., & Gorshkov, V. V. (2004). Revising the fundamentals of ecological knowledge: The biota-environment interaction. *Ecological Complexity, 1*(1), 17–36.

Green, N. P. O., Stout, G. W., & Taylor, D. J. (1984). In R. Soper (Ed.), *Biological science* (Vol. 2). Cambridge: Cambridge University Press.

Holmen, K. (1992). The global carbon cycle. In S. S. Butcher, R. J. Charlson, & G. H. Orians, G. V. Wolfe (Eds.), Global biogeochemical cycles (pp. 239–262). London: Academic Press.

Houghton, R. A., Hobbie, E., Melillo, J. M., et al. (1983). Changes in the content of terrestrial biota and soils between 1860 and 1980: Net release of CO2 to the atmosphere. *Ecological Monographs, 53*, 235–262.

Houghton, R. A., Boone, R. D., Fruci, J. R., et al. (1987). The flux of carbon from terrestrial ecosystems to the atmosphere in 1980 due to changes in land use: Geographic distribution of the global flux. *Tellus, 39B*, 122–139.

Jablonsky, D. (1994). Extinctions in the fossil record. *Philosophical Transactions of the Royal Society London B, 344*(1), 11–17.

Lima-de-Faria, A. (1988). *Evolution without selection. Form and function by autoevolution*. Oxford: Elsevier.

Makarieva, A. M., Gorshkov, V. G., & Vil'derer, P. A. (2014). On the scientific analysis of evolution. Energy: Economics, technology. *Ekologia, 9*, 65–70. (in Russian).

Markov, A. V. (2015). *The birth of complexity. Evolutionary biology today: Unexpected discoveries and new questions*. Moscow: Astrel: CORPUS. (in Russian).

Moiseyev, N. N. (1998). Once again on the problem of coevolution. *Ekologia i zhizn, 33*(7). (in Russian).

Neftel, A., Oeschger, H., Schwander, J., Stauffer, B., & Zumbrunn, R. (1982). Ice core sample measurements give atmospheric CO_2 content during the past 40,000 years. *Nature, 295*, 220–223.

Raven, P. H., & Johnson, G. B. (1998). *Understanding biology*. St. Louis: Times Mirror/Mosby College.

Rotty, R. M. (1983). Distribution of and changes in industrial carbon dioxide production. *Journal of Geophysical Research, 88*(C2), 1301–1308.

Schwartzman, D. W., & Volk, T. (1989). Biotic enhancement of weathering and the habitability of Earth. *Nature, 340*, 457–460.

Watts, J. A. (1982). The carbon dioxide questions: Data sampler. In W. C. Clark (Ed.), *Carbon dioxide review*. New York: Clarendon Press.

Whittaker, R. H., & Likens, G. E. (1975). The biosphere and man. In H. Lieth & R. Whittaker (Eds.), *Primary productivity of the biosphere* (pp. 305–328). Berlin: Springer.

Zavarzin, G. A. (2001). The making of the biosphere. *Priroda, 11*, 988–1001. (in Russian).

Zavarzin, G. A. (2007). The anti-market in nature (Musings of a Naturalist). *Vysshee obrazovanie v Rossii, 4*, 123–130.

Part V
Weighing a Scientific Approach

Chapter 13
Foundations of Sustainability in Nature and Society

The problems of sustainability, of sustainable development, that frequently discussed term of the turn of the present century, we have touched on in previous chapters. But now the time has come for us to discuss it in more detail.

What is it? Is it the true guiding star for escaping the global crisis that has overtaken the world? Or is it only the next in a string of media campaigns, not unlike the pronouncements made from high rostra in the early 1960s USSR of how "The current generation of Soviet people will live under communism?" And does anybody really know what sustainable development is? After all, until quite recently, humanity somehow got along fine without it, governed by age-old experience and practice often based, in any case, on the self-interest of one or another political, national or social group as well as on a system of checks and balances. Meanwhile, relations between states, as a rule, were built on temporary treaties and alliances which, however, could easily be violated in the event of a changing balance of power on the political chessboard. It was, in essence, the path of spontaneous development, and it accompanied global ecological sustainability, determined for the time being by a relatively low population on Earth and its weak technological armament.

With the start of the twentieth century, however, the situation fundamentally changed. Man took hold of hitherto unknown sources of energy and made himself capable of influencing his surrounding world on a scale previously unseen. And while before social cataclysms, revolutions and wars had imposed misfortunes primarily of a local nature, though at times sweeping off whole peoples and states, with the appearance of modern weapons of mass destruction, any full-scale nuclear conflict is capable of annihilating all life on Earth, as shown by Russian geophysicist Georgy Golitsyn, American astronomer Carl Sagan and their colleagues. This Golitsyn-Sagan Hypothesis, more famously known as Nuclear Winter, was checked simultaneously on computer models in the Computing Center at the Soviet Academy of Sciences (Moiseyev et al. 1985) and a team of scientists in America. In both cases, computer calculations confirmed the accuracy of the hypothesis[1]. And what had seemed to be innocent technological novelties, such as the refrigerant Freon, patented in 1928 and widely used in refrigerator parts, in half a century began threatening "ozone holes" over the planet's polar areas.

Along with this, the most important achievement in public thought over the past decades has been the understanding that ecological sustainability cannot be viewed independently of its social and economic aspects. After all, against this backdrop of modern humanity's technological armament, even typical corporate selfishness can lead to dangerous and unpredictable consequences. This nearly happened in the Soviet Union in the 1980s, when projects to divert northern rivers, being shoved through with unflagging obstinacy, were closely connected with the Ministry of Water Resources.

In this way, life itself has put humanity in search of a development path that would not destabilize the environment and, what's more, aid the harmonization of social relations endowed with a sense of responsibility for the fate of our common home, Planet Earth. This idea of universal stability in the natural and social environment, a relation to life as a fragile gift that must be held safe to be passed along as our inheritance to the next generation, pressed in human consciousness in the second half of the twentieth century against the drive to reform and reconstitute the world, which at that time had gripped millions on either side of the Iron Curtain. That was when words of sustainable development as an alternative to the previous, nature-destroying course of civilization sounded from the rostrum at the 1992 Earth Summit in Rio de Janeiro.

Characteristically, the economically successful countries put forward the idea of Sustainable Development first, and for good reason. Having long since destroyed the greater part of their own ecosystems, they recognized sooner than most the ecological consequences the rest of the world would incur in an attempt to follow the same path. Therefore, warnings of the exhaustibility of natural resources amidst civilization's continued expansion, as were heard at Rio de Janeiro, bore witness that this problem had become a fact of public knowledge.

As it often happens, the term Sustainable Development, however, had its own backstory. In the mid-twentieth century, a group of scientists and managers studying issues of fishing regulation in Canada used the phrase *sustainable yield*, meaning a system for exploiting fisheries while not exhausting them. To do this, the yearly catch of fish would correspond to the population's ability to reproduce itself. Nearly a century earlier, the same idea, using different terminology and referring to different resources, was put forward by German foresters. Here it also had in mind an analogous system for exploiting forests in such a way that logging did not exceed natural growth and that the wood harvest occurred without loss to nature. Now such a system is called sustainable forestry. Such resource exploitation may continue indefinitely under constant conditions of climate and other factors that do not depend on human activity.

---

[1] Granted, not all modern climatologists, including supporters of the Golytsin-Sagan Hypothesis, find these calculations convincing. This is because the model turned out to be too sensitive to changes in input data, so even small variations lead to materially different results.

But only at the end of the 1980s did this term receive a new hearing. And, thanks to its use in the Brundtland Commission report, *sustainable development* gained a broad and steady scientific coinage. It would be an exaggeration, however, to think that a quarter century on the global community has a clear, crystallized view of the substance of Sustainable Development or is of one mind concerning the path to its practical realization.

In particular, even its very first definition, given in the Brundtland Report, *Our Common Future* (1987), provided ample ground for disagreement. So, for example, in the Report's second chapter it says "Sustainable development is development that meets the needs of the present without compromising the ability of future generations to meet their own needs." (Our Common Future 1987: p. 41). But how do we understand the needs of future generations? Shall we equate them to the current requirements of those who live in developed countries or of those who only aspire to reach that level but cannot be counted among the number of impoverished? And how do we understand the words on development that does not threaten, "the natural systems that support life on Earth: the atmosphere, the waters, the soils, and the living beings" (Our Common Future 1987: p. 42). if we do not clarify the nature and specifics of the threat? That mankind, to one extent or another, has exerted and clearly will continue to exert a negative influence upon the biosphere does not leave the question of a doubt. Not without cause did Friedrich Nietzshe call it a "disease of the Earth." The whole question is how to stop this disruptive influence from surpassing the biota's capacity to compensate.

In short, much in these formulations appears insufficiently developed and lacks an adequate theoretical base. This methodological shortcoming, perhaps inevitable at the stage of acknowledging the problem, has given rise to a multitude of contradictory and quite arbitrary renderings of this understanding, so crucial to modernity. We observe the greatest inconsistency where discussion touches on the fundamental compatibility of sustainability and growth with the characteristic mix-up or even jumbling of these two conceptual categories.

Thus, some authors assert that sustainability and development contradict one another, and so we ought to reject any pairing of the two (Valyansky and Kalyuzhny 2002). Here you could recall that from a philosophical point of view development is a particular instance of a movement, just as a movement is a particular instance of development: a movement toward civil society, a movement toward social equality, etc. And the sustainability of movement (motion) is one of the fundamental concepts of mathematics, going back to Joseph-Louis Lagrange and Simeon Poisson, then further developed by Henri Poincare and Alexandr Lyapunov. As they thought of it, it meant motion which, having started at a point of some predetermined tunnel, never exits beyond the bounds of that tunnel. The motion here is the product of a change, and the sustainability—of invariability, the consistency of some relation or property of the object, maintaining itself despite any change from among a set of the concrete, fixed class of potentially possible changes (Danilov-Danil'yan 2003).

In such a case, the development of civilization, a social group or economic system can be considered sustainable if it maintains a certain invariant, particularly with concern to the system properties on which its survival depends. For civilization as a whole, this invariant is the limit of environmental pressure, beyond which the adaptive capabilities of the biosphere are exhausted and its irreversible degradation begins (more on that in Chap. 14). With concern to another pair of concepts, growth and development, here disagreement partly owes itself to the polysemy of the English verb *to develop*, meaning at once to develop, to improve, to grow and to expand. This seemingly gives credence to the authors who link sustainable development with growth, even if slowed, limited to available resources and not exceeding the limits of natural ecosystems' assimilated capabilities (Jocelyn et al. 1994).

But, one way or another, the great majority of researchers allow for some form of economic growth as part of sustainable development. Growth has long figured as a panacea in the public consciousness. As the book *Limits to Growth the 30-Year Update* says, "Individuals support growth-oriented policies, because they believe growth will give them an ever increasing welfare. Governments seek growth as a remedy for just about every problem. In the rich world, growth is believed to be necessary for employment, upward mobility, and technical advance. In the poor world, growth seems to be the only way out of poverty. Many believe that growth is required to provide the resources necessary for protecting and improving the environment… For these reasons growth has come to be viewed as a cause for celebration" (Meadows et al. 2006: p. 6).

And yet, for all the intertwining of these understandings, there exists between growth and development a sufficiently deep distinction in meaning, including that fixed in linguistic usage.

So, according to the single-language *Merriam-Webster Dictionary, to grow* means to spring up, to increase in size, to have an increasing influence, and, as a transitive verb, to cause to grow or to promote the development of. And here comes the semantic model of *to develop*, first as a transitive verb: to promote the growth of, to expand by a process of growth. The intransitive verb means to go through a process of natural growth or evolution by successive changes, to come into being gradually, to become manifest.

Thus we can see an important mark of distinction, allowing us, to a certain extent, to develop the concepts of growth and development. While growth is a change quantitative in substance, development is structural and qualitative. And, therefore, each of these processes obeys its particular rules and yields dissimilar results, at times radically different. So, for example, the permanently increasing pressure of civilization upon the biosphere, already having reached the limits of its adaptive capabilities and in places even going beyond these limits, is an obvious example of unbridled quantitative growth, disregarding any regulation or limit and, therefore, incurring the most dangerous consequences. But if humanity, as some think, is doomed to incessant growth in one or another modification, then, in that sense, it stands in sharp contrast with the development of the biota.

Indeed, the process of establishing and evolving ecosystems, from which the very concept of sustainability is borrowed, constructs itself upon a very different foundation than the world built by man, and we can characterize this behavior as a phenomenon of development without growth. Take any tropical forest or tundra community—all of these ecological systems arising through evolution have long developed only qualitatively and, under stable climate conditions, never grow in physical size whether by territory or volume. Note that Vladimir Vernadsky calculated the mass of living material by order of magnitude independent of time. Limits to such qualitative development, in all likelihood, do not exist, for which we have in evidence the colossal complexity of the biota. The stimulus for this arises from the biota's constant "dialogue" with the environment, including the search for the most efficient mechanisms of its own regulation and stabilization, and, in case of disruption, a way to restore the environment to the margin of stability.

After a particularly strong and prolonged disruption, though, such as glaciation, this restoration takes the path of evolutionary speciation, i.e. a radical reconstruction of the biota's internal structure that requires hundreds of thousands or even millions of years, and the conditions thus changed set the benchmark for the next stage of evolution. We should not look upon this benchmark, however, as something pre-ordained, since the evolving biota, including the localized communities, "shifts" it in a direction it "finds convenient." Two factors play a part in this. First, such shifts are determined by the potential of the biota and its communities, and, correspondingly, have limits to their potential. Second, reaching a benchmark at each concrete stage of evolution can be thought of as piece work. As each developing species resolves an evolutionary task, a task of a more general character is simultaneously resolved. That includes increasing the overall adaptive potential of the biota, aiding its survival in case of possible catastrophic changes to the abiotic environment.

***

It would seem that evolution and human progress are both founded on the principle of selection, mutual adaptation and the competition of peoples, cultures and civilizations. And, nonetheless, humanity, unlike nature, embodies the sentiment of incessant and ever-accelerating growth, whether demographic, economic or material, the last of which we often equate with progress. But while competitive relations in the biota are one of the means of providing long-term stability, the case of humanity, as a rule, demonstrates just the opposite inclination. Here competitive relations of civilizational subsystems frequently make themselves the greatest source of global unsustainability.

But, is this, humanity's Achilles' heel, linked to some fundamental particularity of our lifestyle? One should think so. And here, first of all, we'd like to call attention to the very way that humans interact with their environment which sharply distinguishes them from all other living things on Earth. Because while all other species conceive and adapt their life activity to the environment, man, alone among the crowd, took a fundamentally different turn, adapting this world to his own needs and wants. "Man is the only creature who refuses to be what he is," Albert Camus wrote in his 1951 book, *The Rebel*, granted, not in the ecological but in the social aspect of our lives. "The problem is to know whether this refusal can only lead to the destruction of himself and others" (Camus 1991, Introduction).

To explain this, we must make the important distinction between the heritable mechanism of sustainability, the basis of which is genetic memory in the biota, and the supra-biological structure of human civilization, where culture rather than genome supplies the memory. It is also worth distinguishing the base section of a culture—its world view and spiritual or moral values—from the complex of practical knowledge and skills, including the technology that humans use.

While the basal composition of a structure changes very slowly, forming the sustainable core of the society, knowledge and practical experience expand ever more determinedly, involuntarily bringing the surrounding world into the process. This in particular holds the key to the incessant, accelerating growth of civilization, incomparable in speed to the evolution of the biota. It is that incompatibility which gave rise to the ecological challenge of our day. After all, in growing its technological power, its physical and financial capital, humanity could not correspondingly increase the productivity of nature's capital, determined by entirely different processes of its own—solar energy coming to Earth, the plant biota's capacity to use it, the speed of biochemical reactions, and so on.

Thus, the warnings sounded at Rio de Janeiro in 1992, until then understood only by a small clique of specialists. They warned that the global ecosystem was truly exhaustible, that the economy must account for the ecological factor and that technological progress far from always provides social progress. This proved an unquestionable intellectual breakthrough, calling attention to the problem from the widest circles of global society. That same year, a group of about 1700 scientists from 70 countries including 102 Nobel Laureates, members of the Union of Concerned Scientists, came forward with the troubling petition, "World Scientist's Warning to Humanity." "Human beings and the natural world are on a collision course...The earth is finite. Its ability to absorb wastes and destructive effluent is finite. Its ability to provide food and energy is finite. Its ability to provide for growing numbers of people is finite. And we are fast approaching many of the earth's limits....No more than one or a few decades remain before the chance to avert the threats we now confront will be lost and the prospects for humanity immeasurably diminished" (World Scientists'... 1992).

It looks as though the idea of sustainable development came forward just when it became a necessity. It is the first serious attempt to find a way out of the civilizational dead end linked to the very foundations of human existence in which material growth has become an end in itself. The fetishization of growth in recent times has come to worry economists more and more. "The economics of growth and its relationship with development, in particular, require radical rethinking. A vast theoretical and empirical literature almost uniformly equates economic growth with development," It says in the UN Human Development Report for 2010. "Its models typically assume that people care only about consumption; its empirical applications concentrate almost exclusively on the effect of policies and institutions on economic growth" (Human Development Report 2010).

But this psychology has set its roots too deep, pulling into its orbit not only the residents of developed countries, but wider sections of third-world populations, including such giants as China, India and Brazil. At its core, this represents a choice of values, before which, perhaps unknowingly, stands twenty-first century humanity. On this choice, ultimately, the success or failure of transition to sustainable development will depend.

References

Brundtland Commission. (1987). *Our common future: Report of the world commission on environment and development.* Retrieved from http://www.un-documents.net/our-common-future.pdf

Camus, A. (1991). *The Rebel* (Anthony Bower, Trans.). New York: Vintage International.

Danilov-Danil'yan, V. I. (2003). Sustainable development (a theoretical-methodological analysis). *Ekonomika I Matematicheskie Metody, 39*(2), 123–135. (in Russian).

Human Development Report. (2010). Retrieved from http://hdr.undp.org/sites/default/files/reports/270/hdr_2010_en_complete_reprint.pdf.

Jocelyn, M., Johnson, D., John, R., III, & Jocelyn, M. (1994). *Making development sustainable.* Washington: The World Bank.

Meadows, D., Randers, J., & Meadows, D. (2006). *The limits to growth: The 30 year update* (pp. 57–61). London: Earthscan.

Moiseyev, N. N., Aleksandrov, V. V., & Tarko, A. M. (1985). *Man and biosphere: Lessons from systems analysis and modeling experiments.* Moscow: Nauka. (in Russian).

Valyansky, S. I., & Kalyuzhny, D. V. (2002). *Civilization's third path, or will Russia save the world?* Moscow: Algoritm. (in Russian).

World Scientists' Warning to Humanity. (1992). Union of concerned scientists. Retrieved from http://www.ucsusa.org/about/1992-world-scientists.html#.WdUCzWhSw2w

Chapter 14
Sustainable Development Within the Norms of the Biosphere's Carrying Capacity

While humans, in the process of their economic activity, constantly destabilize the environment, the biota, from the moment of its appearance, has supported its stability and its sustainability as a necessary condition of survival. At the earliest stages of life on Earth, single-celled prokaryotes carried out this work, forming the platform of the modern biogeochemical machine (Zavarzin 2004). Later on, multicellular organisms took on the same mission. Primarily these were plants and fungi, which, together with protozoans, form the main part of the Earth's biomass, fill the atmosphere with oxygen, swallow up excess carbon dioxide gas and participate in sediment formation. The World Ocean owes much of its leading role in stabilizing the planetary environment to zoo- and phytoplankton. And at a time when more than 60% of land ecosystems have been destroyed, it is the ocean depths with their still only slightly disturbed biota that serve as the main channel (sink) for removal of excess anthropogenic carbon from the atmosphere. However, even the World Ocean is unprepared to bear the mounting man-made burden.

So, according to estimates by John Houghton and his co-authors, the World Ocean and its ecosystems currently swallow up more than half of atmospheric carbon arising from burning fossil fuels. The rest of it accumulates in the atmosphere. Ocean ecosystems also absorb about two-thirds of "excess" carbon formed on land areas destroyed by economic activity, with the preserved territorial ecosystems swallowing up the remaining third (Houghton et al. 1996). In this way, we can plainly see the violation of the closed-loop cycle of this most important nutrient, leading to its gradual accumulation in the atmosphere. Among everything else, this is, without a doubt, fact number one, unquestionably bearing witness that human influence on the biosphere has already passed the acceptable limits, and that we can consider humanity's exit beyond the ecological carrying capacity a done deal.

You have seen the concept of **the biosphere's *carrying capacity*** (also called economic, ecological or assimilating capacity) as the most important limiter of material human activity before. And while there can be no doubt as to its preeminent role in posing the problem of sustainable development, giving us our most important instrument for a quantitative approach, scientific circles have still not come to a single mind

© Springer International Publishing AG, part of Springer Nature 2018
V. I. Danilov-Danil'yan, I. E. Reyf, *The Biosphere and Civilization: In the Throes of a Global Crisis*, https://doi.org/10.1007/978-3-319-67193-2_14

concerning the concept. Biotic Regulation Theory proposes its own line of thinking in an attempt to provide a scientific basis for it. And though this line of thinking has not yet gained widespread acknowledgment, there is no convincing alternative under review, either. We will, therefore, ponder upon it further, all the more because it has been developed in sufficient detail and distinguishes itself through its logical construction.

Let's begin with the fact that humans, like any other species on Earth, exist within the bounds of a particular energy corridor that characterizes their maximum share of overall energy flows in the biota which they can use for their own needs without risk of environmental disruption. Here we are talking about energy already created by plants on land and phytoplankton in the ocean through photosynthesis and stored in the form of organic material, called *primary production*. The yearly magnitude of this organic material created on a given territory has received the name *gross primary production*, 15–70% of whose plant-stored energy is spent on their own growth and respiration (Leith and Whittaker 1975). Thus, only the remaining portion takes part in the further cycle, used by consumer organisms of the next trophic level. That portion represents *net primary production*. A typical example of net primary production would be the yearly falling of leaves, dry branches and seeds at temperate latitudes.

But this is only the tip of the iceberg, because the essence of net primary production flow contained in organic plant material is in the transfer of energy from one group of organisms to another, from one trophic level to the next, and the overall number of levels can reach four, five, or even six. As materials from field research conducted in a large number of different ecosystems has shown, the rule for distributing this flow of energy applies itself strictly and shows itself equally characteristic for the most varied natural communities. In sum, allowing for simplification and rounding, the results establish that **90%** of net primary production in ecosystems goes to use by bacteria and fungi, which also play the role in regulating the environment. **Ten percent** is used by invertebrates (arthropods, worms, molluscs, etc.). With concern to vertebrate animals, they receive less than 1% of energy circulating in the biota (Fig. 14.1).

The demonstrated characteristics show high stability and clearly preserve (or preserved prior to our time) their values within a narrow interval of possible variation over a period of tens of millions of years. Biotic regulation theory factually equates the universality of this distribution to an ecological law (Gorshkov 1981; Gorshkov et al. 2000). In this way, the 1% energy corridor for large animal species developed through evolution should be viewed, according to Gorshkov, as a kind of defense mechanism, protecting the biota from chance fluctuations arising in flows of organic material synthesis and decomposition.

However, the question is fair: how much variation from this parameter value is acceptable for the system? Let's go back to the drawing in Chap. 10 (Fig. 10.2) where we discussed the nature of climatic sustainability maintained by the biota. The illustration presents this sustainability in the form of a symbolic ball, located in a "climate hollow." The insignificance of the systems divergence from input parameters testifies to the perfection of the regulatory mechanism. But since we are not dealing with a scalar value or a variation interval but with a region of n-dimensional space in which the quantity value of *n* is unknown to us, the value of a given parameter is less important to us than the "construction" of the climatic hollow, i.e., the ball's area of sus-

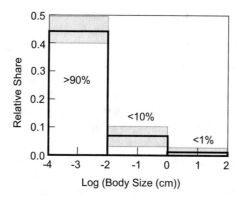

Fig. 14.1 Distribution of organic material decomposition (destruction) speed by organism body size (1) among those living on land. (1) bacteria and fungi, (2) invertebrates, (3) larger animals (starting from rodents). The solid line represents the universal distribution of organic material destruction observed in undisturbed ecosystems. Percentages correspond to the share of use of net primary production by each of the three groups of organisms. The area under the solid descending line is equal to unit. The thatched portion represents variation from mean distribution in separate ecosystems Source: Gorshkov et al. (2000, p.79)

tainability and its "size." However, Biotic Regulation Theory does not encompass this type of question, reducing the whole problem to a single variable, energy, and through its silence supposes that this persistence of parameter is an inseparable quality of the defense mechanism of Earth's biota. But such a presupposition requires, in our opinion, a corresponding basis which, unfortunately, the authors have left out.

***

As we have said, the most important task of the biota according to Gorshkov's conception is, at both the global and local level, supporting the high level of impermeability in the cycle of matter, i.e. the maximal parity between the processes of organic matter synthesis and decomposition, without which the environment would quickly degrade to a state unsuitable for life. This task, Gorshkov claims, is decided by the efforts of the great multitude of mutually uncoordinated autonomous individuals (organisms) which make up the living membrane of any ecosystem. He supposes that the extant type of internal correlation among living individuals in the population stabilizes itself on the basis of competitive interaction and that selection is executed only if all individuals in the population are completely independent and mutually uncoordinated. Otherwise, the removal of a defective individual from the population would be completely impossible, just as it is impossible to remove a diseased organ from an organism (Gorshkov et al. 2000). Only under this condition does it become possible to minimize the great many chance fluctuations that threaten the survival of any complex organized system.

Indeed, according to Gorshkov, bacteria and saprobiontic fungi, responsible for 90% of organic decomposition, are either self-sustained organisms (bacterial cells) or weakly-coordinated multicellular structures (fungi) made up of filamentous structures—spores—a few microns in breadth, which can be found by naked eye or with the aid of a magnifying lens. The vegetative body—the fungal mycelium—looks like a web, or a fluffy, velveteen frost, or even a thin film. On the forest floor, these fungal filaments can reach 35 km for each gram of soil cover.

Something similar occurs among plant-producers. They are also multicellular structures made up of weakly coordinated parts (modules) for which even the lifespan may differ. So at mid latitudes, the leaves of trees retain viability for the course of a mere season, which cannot be said of the trees' roots and trunks. Thus the death of some part of a plant, such as when herbivores eat its branches, does not lead to the death of the whole organism and even stimulates the development of its other parts, a possibility absolutely excluded from the strictly coordinated bodies of animals. At the same time, leaves of the very same tree may compete with one another for sunlight and nutrition (Gorshkov et al. 2000).

In this way, plants, fungi and bacteria have all reached a balance between organic synthesis and organic destruction by fundamentally the same route. Environmental complications arise only once you introduce a "disturber of the peace" to the ecosystem—vertebrate animals.

So, if you take consumers of the second order—rodents, lagomorphs, ungulates, primates, most birds that feed on plant biomass—then, as calculations show, their metabolic power by projection area surpasses that of plant productive power by several orders of magnitude. And if you compare human metabolic power (about 150 W at a body projection area on the order of 0.5 m^2) with the average power of photosynthesis (0.1 W/m^2), the difference in energy flows used as calculated by surface area unit adds up to more than three thousand times (Makarieva et al. 2014).

Therefore, animals, depending on body size, need to consume food synthesized by plants over a territory hundreds or thousands of times greater than their projected area, and also eat in mere hours what plants took a year to synthesize. As a result, animals are forced to constantly move about their feeding grounds, which is a necessary condition of their survival. At the same time, animals devouring accumulated plant production inevitably leads to its sharp variation since the consequent restoration of biomass occurs at an entirely different rate. This variation, in turn, overlaps with variations in excrement in the environment left by animals after food ingestion, which also leads to violation of its stationary state.

Therefore, according to Gorshkov, the existence of large animals, from mice to elephant's is possible in conditions of a highly closed matter cycle only with a minimization of their disruptive influence on the ecosystem, according to which the average usage quota for plant production must not go far beyond the limits of natural fluctuation. And since the variation of consumed biomass grows along with the size of the animal, we should see parallel to this a corresponding reduction in the quota accruing to the species in accordance with observed distribution of use of net primary production by organism body size (Gorshkov et al. 2000, p. 104). And the biota resolves all the tasks through its own specific methods in each case.

So, if the population density of some herbivore species grows too dense, among plants the share may increase of breeds that have thorns or possess a taste repulsive to animals. We all know of plants that influence the numbers of one or another species by means of medicinal, or, on the contrary, poisonous or narcotic substances. Meanwhile, the community that best reacts to variations in large animal numbers in one of the above-mentioned ways receives an advantage over its neighbors whose ability is more weakly expressed, and in the process of competitive struggle will come out the winner.

In this way, the theory follows, in relation to a biotic community developed by plants, fungi and micro-organisms, large herbivorous animals are just as much a component of the environment regulated by them, as well as the nutrient elements contained in the soil and air, whose stable concentration the biota maintains century after century.

With concern to predators, at the top of the ecological pyramid, they cannot exceed their optimal numbers while maintaining stable numbers of prey. Therefore, under natural conditions, predators cannot violate ecological equilibrium, and their function in the ecosystem is founded on removing defective individuals with altered heritable programs from the population and reducing it to equilibrium state. And, clearly by no coincidence, the growth of polymorphism in herbivores that arises from serious disruption of habitat and decay of the associated genetic program, is, as a rule, accompanied simultaneously by parasites and predators. We observe an analogous correlation between plants and plant-eating insects, where the increase in such polymorphism in the former is accompanied by population growth in the latter. Situations like this arise after forest fires, clear cutting or other major disruptions to the natural environment (Isaiev et al. 2001).

And, so, when large animals truly represent a danger of destruction to biotic communities, it is primarily conditioned upon them exceeding a certain critical limit to population growth. Therefore, normal behavior in higher animals with a pre-served heritable program is usually directed at maintaining a stable population density. Both because of limited birth rate during food shortages, as shown in the example of tundra wolves (Chap. 2), and due to control over feeding grounds by animals themselves through the use of, for example, sound signals warning that territory is occupied (McNab 1983), etc. Migration of animals at times of excess population density serves the same purpose, as does activation of parasites and predators, aiding in the reduction of herbivore populations. Such intra- and interspecies interactions, in Gorshkov's opinion, are absolutely necessary to provide competitive advantage to biotic communities. After all, ultimately it is not species that survive, but communities (if, of course, you accept the axiom that intercommunity competition truly represents the dominant type of relationship).

But why does the biosphere need animals at all, if it got along just fine without them over the course of hundreds of millions of years? And even today, the share of energy flows going to them is so low that they cannot play a noticeable role in the biosphere's overall energy system. Nonetheless, the universal spread of large animals bespeaks a place occupied in the biosphere clearly necessary for ecosystems. They, too aid in the maintenance of environmental stability, though one must look beneath the surface to explain this contribution. This is how Viktor Gorshkov approached a solution to this problem, lumping large animals together with "reconstructive" plant species.

As we said in Chap. 12, these species play an important but specific role in the process of succession, shifting the concentration of food substances in the environment toward a direction disadvantageous for themselves but advantageous to the next generation, paving the road, it would seem, to the climax community's rebirth. External physical disruption of the biota, however, whether by fire, volcanic eruption, hurricane or other meteorological extreme, takes on a randomized and irregular character, and, under conditions of extended preservation of the climax phase, reconstructive species required for the post-disruption stage are gradually pushed out of the ecosystem. At this time, they exist as isolated individuals, making up a disparate population to which the mechanisms of competition and selection practically do not function. This, in turn, incurs the decay of their genetic programming.

Therefore, under the threat of reconstructive species degrading and disappearing, the biota should have a mechanism for regular disruption of ecosystems, which would support the "reconstruction" population at its minimum necessary level. And here, clearly, large animals carry this function out, playing the role of persistent ecosystem disruptors, independent of the major external disruptions. Bringing destruction to vegetative cover, they create beneficial conditions for the survival of reconstructive plant species that multiply in number upon the zone of destruction (Gorshkov et al. 2004).

In this light, it becomes clear not only what the ecological mission of ambulatory animals is, but why they occupy so humble a place in the biota's overall energy system, where inanimate organisms—plants, bacteria, fungi—play the main role. If you were to cut off large animals and birds from the whole mass of organisms dwelling on Earth, you could compare the biosphere to an energy generator providing that group of species with the energy necessary for life at an energy conversion efficiency of no higher than 1%. The other 99% of the biosphere's energy power goes toward supporting environmental stability (Gorshkov et al. 2000, pp. 104–105). Such, according to Biotic Regulation Theory, is the biospheric energy structure developed through evolution, allowing the maintenance of a highly closed matter cycle as flows of organic synthesis and decomposition coincide with exactness on the order of 10^{-4} (see Chap. 12).

Today, however, the extent to which this biochemical cycle is closed has lowered by nearly an order of magnitude as is visibly demonstrated on the bar chart below (Fig. 14.2). At first glance, it differs little from what was presented in Fig. 14.1, as long as you don't count the added dotted line. But this chart characterizes the current disrupted state of the biosphere, directly linked to the economic activity of humans. And we are allowed to judge the quantitative side of this by the biosphere's carrying capacity—that unit of the acceptable extreme of human civilization's influence on the environment on which, in Martin Holdgate's phrase, many ecologists first cut their intellectual teeth.

Humans, after all, also belong to the category of large animals. And, if you follow Biotic Regulation Theory, the same ecological limits listed above apply to them as well as to the animals they have domesticated. And in order not to cut the branch we are sitting on from under us, humans and all of their business ought to fit themselves into the bounds of the energy corridor that the biosphere has assigned to all large animals. Granted, given the current level of knowledge, we can only speak of the size of this "corridor" with certain caveats. Thus, we should more likely speak of an order of magnitude, perhaps close to 1%, but nonetheless differing from it. To put it mathematically, the number is $1 + \varepsilon\%$, in which the value ε has not yet been determined by science.

But let us take Gorshkov's lead with $\varepsilon = 0$. Then, based on that assumption, the size of the biospheric corridor humans should fit into along with other large animals comes to an order of magnitude of 1%. This assigned 1% energy quota that humans can access without undermining environmental sustainability is most conveniently expressed in scale of net primary production. Its size could be expressed in organic carbon mass

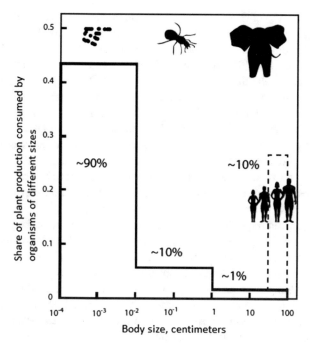

Fig. 14.2 Share of organic material consumption on land by size of organism (bacteria and fungi, invertebrates, large animals), taking into account anthropogenic disruption of the biosphere. Solid line—universal distribution observed in all undisrupted land ecosystems. Area under the solid lines is equal to 1 (100%). Figures in percentages correspond to the share of consumption by each of the three organism groups. The dotted line characterized modern anthropogenic disruption of the land biota. Area under the anthropogenic column corresponds to food for humanity and livestock, as well as forest consumption. Source: Biotic Regulation Website: http://www.bioticregulation.ru/life/life2_r-6.php

(tons) or in terms of power (watts) adequate for the quantity of biomass produced by plants on a given territory in a year beyond that expended on growth and respiration by the plants themselves. And if the energy power of the whole land biota has an order of 100 terawatts (TW, 10^{12} W), then 1% of that would be equal to ~1–2 TW. Based on a valuation of summary mass of synthesized organic carbon (~10^2 GtC/year), we receive an order of magnitude of 1.0 GtC. Therefore, **1–2 TW** (in power units) or **1.0 GtC** (in units of organic carbon mass) gives us a quantitative idea of the biosphere's carrying capacity—the maximum size, beyond whose limits human civilization must not go if concerned for maintaining stability of the global environment.

Today, however, we have already surpassed that limit by an order of magnitude. One-two terawatts corresponded to civilization's power at the start of the previous century when, Gorshkov supposes, it stepped across the forbidden boundary. And it is clearly no coincidence that the rapid increase in atmospheric CO_2 also began around 1900, after which humanity went over the limits of the 1% energy corridor. At that time, Earth's population added up to 1.6 billion people and had already destroyed or seriously deformed ecosystems over 20% of Earth's territorial surface. In this way, accounting for then existing technology and assuming that E = 0, the "geographic equivalent" of the biosphere's disruption threshold might be considered 20% economic integration of land area.[1]

But what does this 20%, this one-fifth of the land's surface that we have long passed by, mean if the dizzying gallop of Twentieth Century has overstepped this boundary three times over and the area of destroyed ecosystems has now surpassed 60%? This is also an indicator of extreme disruption in the biosphere, whose compensatory capacities, clearly, are close to exhaustion. The violation of the closed-loop nutrient cycle shows this (CO_2, nitrogen and phosphorous compounds) as does the progressive loss of biodiversity. The transition of many recently renewable natural resources, most of all water, into the pool of unrenewable or only partially renewable resources clearly demonstrates this as well. So do many other signs, of which we spoke in Chap. 1. All of these troubling symptoms demand the most serious attentions, even regardless of how one handles this theory or that.

But if this first critical boundary for civilization has already been crossed, that opens up the question of the next, far more dangerous threshold, when environmental degradation becomes irreversible and the biosphere loses its capacity for regeneration for an indeterminably long, even on a geological scale, period of time. And here we'd like to turn our attention to the claims of so-called "technological optimists" putting their faith in the unlimited possibilities of scientific and technological progress which has more than once pulled humanity back from the brink and must, therefore, have a handle on the current ecological threat. Because, in light of the biosphere's limits of mass and energy, hopes for artificial environmental regulatory mechanisms destined someday to replace the natural mechanisms look especially groundless.

Indeed, who could doubt that humans have a very long way to go before they learn to regulate and manage the environment with the same energy conversion efficiency and at the same energy level accessible to the biosphere, even if you assume humans capable of mastering such technology at all? The biosphere itself came to it through a multi-billion year process of evolution. And in order to more clearly imagine mankind's abilities as far as creating an artificial environment is concerned, let us again recall the basic "expense account" of the total energy budget our Earth has at its disposal thanks to the solar radiation it receives.

The overall power of this radiation on the boundary of Earth's atmosphere adds up to 10.5×10^6 kJ/m^2 per year. Of that quantity, about 40% is immediately reflected by clouds, atmospheric dust, ice cover and mountaintops, and another 23% is swallowed up by the atmosphere, transforming into heat energy or expended on water evaporation. In this way, the Earth's surface and its vegetative cover is reached by just half of the original solar radiation, or about 5×10^6 kJ/m^2 per year (the actual amount of energy in a given place depends on its geographic latitude). From this

---

[1] Notably, human rights advocate and member of the Soviet Academy Andrey Sakharov came to a very similar assessment in 1974, well recognizing the link between the preservation of the natural environment and destructive human activity. In his article, "The World in Half a Century," it says that to provide a sustainable biospheric balance for the future, it is necessary to divide land into settled and little-inhabited land at a ratio of 3:8 (Sakharov 1990).

half, however, only 25% of light energy has the wavelength suitable for photosynthesis, and only about 0.4% of such rays are used by plants for pure biomass increase, which is roughly 1% of the energy that gets to plants (Green et al. 1984, vol. 1 Ch. 9.2.1, vol. 2 Ch. 12.3.4). It is this insignificant share of the sum total of solar energy that gives rise to energy flows in the biota, whose total energy power, 100 TW, allows it to maintain stability of temperature, climate and other environmental parameters.

By the way, theoretically, this is still not the limit and the biota could, in principle, increase its power by an order (of magnitude), for example, by accounting for plants of the C_4 group, which synthesize carbon based on the tetracarbon acid cycle (Govindjee 1982). These include, in part, corn and sugar cane. However, as calculations have shown, the biota's current power is at the biological limit of sustainability of the current climate, beyond which unpredictable surface temperature fluctuations must surely follow (Gorshvov 1990). Therefore, even if we suppose that humans will someday have an unlimited source of "ecologically pure" energy (cold fusion, solar cell installations in space, etc.) and are able to take management of the environment in hand to provide the same closed-loop matter cycle with the same energy conversion efficiency as the modern biosphere, they would still not be able to go beyond the limits of the biosphere's current power without risk of permanently unbalancing the climate. And, meanwhile, 99% of all energy expended by civilization would need to be spent on maintaining environmental stability. (After all, even today the cost of efficient purification equipment can reach half the cost of an operating business.)

So, what would then be left to satisfy our own wants and needs? About the same and even less than we could have at our disposal with a natural biosphere without expending a single kilowatt on maintaining environmental stability or even thinking about how to deal with the task of a living biota. And now, tell us, is there even the slightest shred of truth behind the ruminations that humans could someday get by without nature?

References

Gorshkov, V. G. (1981). Distribution of energy flow among organisms of different sizes. *Zhurnal obshchey biologii, 42*, 417–429. [in Russian].

Gorshvov, V. G. (1990). Biospheric energetics and environmental stability. In *Itogi nauki i tekhniki, Teoreticheskie i obshchie voprosy geografii* (Vol. 7). Moscow: VINITI. [in Russian].

Gorshkov, V. G., Gorshkov, V. V., & Makarieva, A. M. (2000). *Biotic regulation of the environment: Key issue of global change* (p. 367). London: Springer.

Gorshkov, V. G., Makarieva, A. M., & Gorshkov, V. V. (2004). Revising the fundamentals of ecological knowledge: The biota-environment interaction. *Ecological Complexity, 1*, 17–36.

Govindjee (Ed.). (1982). *Photosynthesis, Development, carbon metabolism and plant productivity* (Vol. 2). New York: Academic Press.

Green, N. P. O., Stout, G. W., Taylor, D. J. (1984). In R. Soper (Ed.), Biological science (Vols. 1 and 2). Cambridge: Cambridge University Press.

Houghton, J. T., Meira Filho, L. G., Callander, B. A., Harris, N., Kattenberg, A., & Maskels, K. (Eds.). (1996). *Climate change 1995: The science of climate change*. Cambridge: Cambridge University Press.

Isaiev, A. S., Khlebopros, R. G., Nedorezov, L. V., Kondrakov, Y. P., & Kiseliov, V. V. (2001). *The dynamics of insect population numbers*. Moscow: Nauka. [in Russian].

Leith, H., & Whittaker, R. H. (1975). *Primary productivity of the biosphere*. New York: Springer.

Makarieva, A. M., Gorshkov, V. G., & Vil'derer, P. A. (2014). On the scientific analysis of evolution. Energy: Economics, technology. *Ekologia*, (9), 65–70. [in Russian].

McNab, B. K. (1983). Energetics, body size and the limits to endothermy. *Journal of Zoology, 199*(1), 1–29.

Sakharov, A.D. (1990). *The Peace, Progress and The Human Rights. Articles and speeches*. Leningrad: Sovetsky pisatel'. [in Russian].

Zavarzin, G. A. (2004). *Lectures on environmental microbiology*. Moscow: Nauka. [in Russian].

Chapter 15
Prerequisites for Sustainable Development and Maintaining Ecosystems by Country and Continent. Russia's "Special Project"

It would likely be no exaggeration to say that our planet, from poles to equator, has already long been suffocating in the human embrace known as anthropogenic pressure. Notably, however, not only people distant from science but many specialized ecologists have not yet recognized the central point of the current global problem. It isn't air pollution in major cities, from which millions of people suffer, and not industrial runoff poisoning our rivers. It isn't even climate change, the anthropogenic share of which might still be debated. The main ecological result of human economic activity, as we have tried to show throughout this book, is the destruction of ecosystems over enormous swaths of land, as well as the water areas of semi-enclosed seas and offshore ocean zones. This sharp weakening of the biota's functions to form and stabilize the environment threatens the biosphere with the most catastrophic consequences. And only an orientation toward natural strengths, toward the preserved biotic potential could, perhaps, prevent the worst possible outcome of these unfolding events.

And if the primary task of sustainable development is to lessen anthropogenic pressure to a level corresponding to the biosphere's carrying capacity, we should, therefore, speak not only of ceasing every variety of "assault" upon nature, but as was said in *Beyond the Limits*, "drawing back, easing down, healing" (Meadows et al. 1992, p. xv). And this means not at all a metaphorical retreat, but an entirely real one in the form of humans emancipating part of the territory they have conquered, as necessary for the biota to perform its planetary mission.

It's probably not worth mentioning how complex and unprecedented this task is, for the resolution of which, in the phrase of Russian Academy member Nikita Moiseyev, humanity will have to walk along a razor's edge. By the way, it would be just as fair to say through the eye of a needle, particularly taking into account the diversity and inequality of starting conditions the various countries find themselves in today in terms of their transition to sustainable development. It is enough, for example, to compare some states in Asia and Africa with all their endemic traits of late feudalism with the countries of Western Europe and North America, truly having reached an information society, to understand the whole depth of the cultural and socio-economic rift which complicates the already difficult problems before the

© Springer International Publishing AG, part of Springer Nature 2018
V. I. Danilov-Danil'yan, I. E. Reyf, *The Biosphere and Civilization: In the Throes of a Global Crisis*, https://doi.org/10.1007/978-3-319-67193-2_15

global community. Add to this also examples of striking divergence in mentality, national traditions and religious beliefs. How then, you might ask, do we reduce this to the common denominator on whose basis alone we can come to a general consensus in developing a global sustainable development strategy?

Yet there exists, at least, one common criteria for all that allows us to compare and contrast the countries of the world in the aspect that interests us, regardless of their social and cultural particularities, their industrial infrastructure development or their mineral wealth. That is the extent to which their natural ecosystems are preserved. That, too, is wealth in the long term—a wealth far more substantive than diamond veins or gold ingots in a bank vault. Only it is a wealth thus far not totally appraised. And if we see the preservation and rebirth of wild nature areas as one of the goals of sustainable development, that means we must consider countries where such nature remains whole to be the stewards of our common patrimony. By the same token, the countries whose territory has lost all or nearly all of its natural ecosystems are "ecological debtors" to the biosphere, even if their environment (as often happened in third world countries) suffered as a result of exploitation by others, including industrialized states. From this position, we will try to assess their starting potential for transition to sustainable development focusing primarily on social and natural parameters and temporarily disengaging ourselves from the others.

In order to put together an image of ecosystem destruction by country and continent, it is best to turn to satellite data. This presents us with a reliable image of the state of the biosphere. Granted, the estimates used differ, which is natural. But most of them still coalesce around the relative value of 60% (totally or partially utilized portion of land) to 40% (unutilized portion). The most notable research was carried out by Lee Hannah and his coauthors (1994).

According to their published data, Earth retains 39.5% undestroyed and 24% partially destroyed ecosystems, occupying altogether 94 million km². Undestroyed ecosystems are characterized by the presence of natural vegetative cover and low population density (less than ten people per km²). Partially destroyed ecosystems pertain to territories on which temporary or permanent agricultural lands abut secondary but naturally restored vegetation and traces of human activity are observed: logging, livestock grazing in which density exceeds pasture restoration capacity, etc. If you throw out ice-covered territory and mountain heights such as Antarctica, Greenland or the Himalayas which have zero biological productivity, the area occupied by undestroyed and partially destroyed ecosystems falls to 52 million km² (Maksakovskiy 2008, Book 1). These ecosystems, however, are distributed very unevenly across the land's surface.

So, along with islands of wild nature left whole from 0.1 to one million km² in size, we also see preserved enormous territorial masses of many millions of square kilometers, spread principally within the bounds of Earth's two main forest belts— North and South.

The first of these, occupying an area of twenty million km² between 45° and 70° north latitude, takes up most of Siberia and the Russian Far East besides its southern regions, the north of European Russia and Scandinavian nations, as well as the northern part of Canada and Alaska. This is mainly boreal forest specific to the cold zone, made up of two-thirds coniferous breeds, and it occupies 38% of all forest-covered land. Roughly half of boreal forests belong to undestroyed forest ecosystems, having so far suffered insignificant anthropogenic influence (Olsson 2009).

The southern forest belt is made up of tropical rainforest in the equatorial and subequatorial zones between 25° north latitude and 30° south latitude. It also occupies about 20 million km^2. The greatest mass of tropical rainforest is spread across South America (the Amazon Basin), Southeast Asia, the islands of Oceania and Africa (the Congo Basin and the area around the Gulf of Guinea). Almost half of all species dwell in these rain forest areas, where more than half of the Earth's phytomass is located. Trees on a single acre grow several times faster than in the northern belt, so the southern belt creates 70% of all net primary production (Forest Encyclopedia 1986). Along with this, due to year-round high temperatures and moisture, organic matter decomposes here very quickly, as a result of which tropical forests deposit just over half as much carbon as boreal forests.

Periodicals have long resorted to the trite if not entirely fitting cliché—comparing forests to the Earth's lungs. We might more justly call them Earth's kidneys, since they filter out excess quantities of nutrients and remove them from circulation, carbon dioxide first among them. Some authors even refer to soil humus and peat bogs as "eternal" carbon sinks, where under corresponding temperature conditions it might stay, as it does embedded in the ocean floor, for an interminably long time (Vompersky 1994).

But all of this holds true only for undisturbed ecosystems, including for example, the virgin climax forests that carry primary responsibility for environmental stability. Disturbed ecosystems, such as forests that undergo logging or cutting back, behave entirely differently. So, on territories utilized by humans, as research data shows, the biota not only fails to swallow up excess atmospheric carbon, it serves as a source of emissions itself. And reserves of carbon accumulated in such forests, according to data from the UN Food and Agriculture Organization, fall at a rate of 1.1 gigatons per year (Global Forest Resources Assessment 2005).

A multitude of samples, taken from 1958 at various observatories around the world, testify to the unflagging growth of CO_2 concentrations in the atmosphere (Fig. 15.1). Analysis of the gas makeup of air bubbles in Antarctic ice cores allows us to form a conception of trends in its atmospheric concentrations before and after the global disruption of the biosphere coinciding with the Industrial Revolution (late eighteenth-early nineteenth century) (Friedli et al. 1986; Staffelbach et al. 1991; Raynaud et al. 1993). As the research shows, deposited CO_2 concentrations adding up to 280 parts per million (ppm) and remaining nearly unchanged over the course of several millennia have now reached 340 ppm, i.e. 28% higher than the preindustrial level (Lorius and Oescher 1994). This increase began before wide-scale use of fossil fuels and overlaps with carbon emissions caused by land usage. From that time to the end of the nineteenth century, the biosphere maintained its sustainability mainly through the efforts of the little-disrupted ecosystems of the World Ocean, the compensatory capacities of which reached their limits at the start of the twentieth century. Thereafter began a process of global change to the environment.

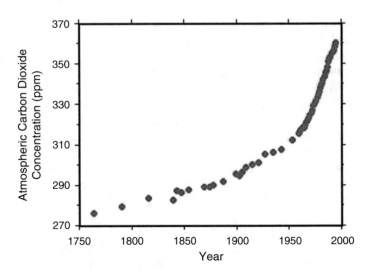

Fig. 15.1 Growth in world carbon dioxide emissions as a result of organic fuel combustion. Source: Worldwatch Database (2000)

In recent decades, several attempts have been made to calculate the balance of atmospheric CO_2 on the basis of the law of conservation of mass and the stoichiometric (volume-mass) ratio of O_2/CO_2 for the main pools of anthropogenic carbon sources and sinks. One of the best known belongs to John Houghton and his co-authors (1996). This research laid the groundwork for the understanding that changes to carbon content occur for communicating media—the atmosphere, the ocean, land biosphere and the hydrocarbon fuel deposits added to them over the past two centuries. The total of these combined flows should, in principle, be equal to zero, i.e. sources of carbon dioxide emissions should be compensated by sinks.

Besides hydrocarbon fuel, combusted in gasoline engines and the furnaces of power plants, large quantities of carbon dioxide enter the atmosphere in the process of cement making, burning of associated petroleum gas and as a result of agricultural activity where biomass is destroyed (clear-cutting forests, destruction of soil humus in tilling, etc.). In turn, the atmosphere, World Ocean and undestroyed ecosystems serve as sinks, where carbon dioxide gas can go and accumulate.

The scale of CO_2 emissions from fossil fuel combustion since the beginning of the industrial era has been studied quite well. At present, when recalculated as carbon, it is estimated to be 5.9 plus or minus 0.5 gigatons of carbon per year. We also know the speed at which CO_2 accumulates in the atmosphere—on the order of 2.2 GtC/year. Finally, from the ratio of $^{13}C/^{12}C$ isotopes in ocean water and air, we can determine the rate at which the ocean absorbs carbon dioxide gas accounting for physical and chemical processes. Excess CO_2 diffuses through the surface at the air-water divide, evening out its concentration according to Henry's Law. It's estimated to be 2.6 GtC/year (Zalikhanov et al. 2006).

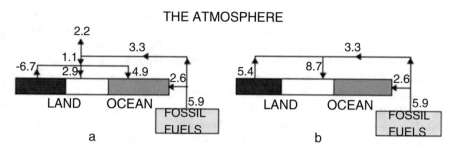

Fig. 15.2 Global carbon flows: (**a**) under current biospheric conditions; (**b**) in the event of partial restoration of forest-land ecosystems, enabling us to stop the process of atmospheric CO_2 accumulation at existing levels of fossil fuel combustion. White boxes—unutilized, virgin biota; black—economically utilized portion of land. Relative size of boxes corresponds to their share of net primary production out of the biosphere's whole production: (**a**) World Ocean—40%; utilized land—30%; unutilized—24%; (**b**) World Ocean—40%; utilized land—29%; unutilized 31%. Numbers on arrows—carbon flows in GtC/year (Gorshkov et al. 2000, p. 170)

The business of evaluating carbon emissions as a result of land usage is a bit more complicated. Here we must determine the reduction in biomass on a given territory and its absorption by undestroyed land and ocean ecosystems. The lack of exact methods to calculate production and destruction of organic material presents a particular difficulty which does not always allow us to avoid mistakes. So, for example, carbon might come out of the territorial biota as a result of ecosystem elimination, primarily of the forest type, and destruction of soil cover. But while, after its exploitation is finished and it has not gone too far, a forest retains its capacity to restore plant biomass and absorb excess CO_2 accumulating in the atmosphere, we cannot say the same of soil, land's greatest reservoir of nutrients. Soil gathers carbon extremely slowly, and loses it very fast. Thus average carbon losses in soil during cultivation add up to 30%, sometimes reaching 70% in the tropics. That means that alongside restoration of damaged ecosystems, as a rule, we observe a reduction in soil carbon content, and as a result, the compensation proves incomplete (Vitousek et al. 1986; Wofsy et al. 1993).

Here we present a precise balance of the global carbon cycle, recalculated by Viktor Gorshkov, Kirill Kondratyev and their coauthors (Gorshkov 1998; Gorshkov et al. 2000, pp. 168-171) accounting for the "biotic pump" effect of the ocean and the contribution made by disrupted and undisrupted land ecosystems (Fig. 15.2a). It is based on data for yearly flows of net primary production in the biosphere as a whole (on order of 100 GtC/year) as well as the World Ocean separately (40 GtC/year) and land (60 GtC/year). The latter, in turn, is divided into two unequal parts—economically utilized land, which makes up 36% of the biosphere's overall flow of net primary production, and the undisrupted biota, making up 24%. But while undisturbed ecosystems, in accordance with Le Chatelier's Principle, bind carbon accumulated in the atmosphere by way of organic synthesis, disrupted ones, on the contrary, themselves give rise to its emission.

Another source of global CO_2 emissions, which the majority of authors tradition-ally assign the leading role, is organic fossil fuels. And here there is no unanimity of opinion among scholars concerning the fate of "lost" carbon that comes up in attempts to settle the balance between CO_2 emissions formed from fossil fuel com-bustion (5.9 GtC/year) and CO_2 accumulated in the atmosphere (2.2 GtC/year), as well as carbon dioxide swallowed up by the World Ocean accounting for processes of physiochemical absorption (2.6 GtC/year). Meanwhile, these scholars far too often underestimate the roles of the ocean biota and land ecosystems, which, in our view, is deeply mistaken. First of all, because in long-passed geological epochs it was precisely through the mediation of the oceanic "biotic pump" that enormous quanti-ties were removed from the atmosphere and buried in the sea floor, to which paleon-tological data bears witness. Second of all, because undisrupted land ecosystems with their soil humus, tundra bogs and boreal forest turf in their long-term (in the opinion of some authors, "eternal") role as carbon reservoirs, truly "model" analo-gous ocean systems with their floor deposits and little-mixed cold waters of the deep.

Artificial agrocenoses, however, such as harvested forest, tilled land, pasture, etc., behave completely differently. Permeation of the biological cycle goes above 10%. As we said in Chap. 12, carbon emissions from disrupted ecosystems match the order of magnitude of those from burning coal, oil and gas (Watts 1982; Houghton 1989) and, according to Gorshkov's calculations, even go beyond them (−6.7 GtC/year). By the way, for these emissions, the absorptive power of the little-disturbed ocean biota (4.9 GtC/year) and undisturbed land ecosystems (2.9 GtC/year) suffices for now. With regard to the problem of "lost carbon," according to the same calcula-tions, it is solved as follows. Of the 5.9 GtC formed by fossil fuel combustion each year, 2.6 GtC/year dilutes into the ocean as a result of physiochemical absorption, 1.1 GtC/year is swallowed up by the ocean biota and the remaining land ecosystems, and the other 2.2 GtC/year accumulates in the atmosphere (Fig. 15.2a).

This accumulation of CO_2, incurring the growth of the greenhouse effect and climate destabilization, represents one of the central problems of modern civiliza-tion and must be stopped by any means necessary. For this purpose, humanity has two paths. The first of them is a rejection of fossil fuel use, which but 20 years ago seemed totally unthinkable. The unbelievable progress in renewable electric energy production, however, has radically changed the situation, and plans to reduce fossil fuel combustion by five-ten times by 2050 no longer look utopian.

But there is another opinion. Among its supporters is Viktor Gorshkov, who con-siders the primary root of this evil to be expanding energy usage itself. What follows is the crux of his argument. If humans had not succeeded so well in exterminating wild nature and anthropogenic disruption of the land biota were much lower than the threshold of destruction, then absorption of fossil carbon by the land and ocean biota would entirely compensate its emissions. At the same time, a transition to ecologically clean energy sources would not, in his view, solve the problem, since disruption to the global biota is determined by the scale of energy usage rather than its source. Thus, for example, the installation of a large number of solar generators in the desert should, in accordance with the Stefan-Boltzmann law,[1] lead to an increase in surface temperature caused by the transformation of light energy into heat. Diverting the expended heat energy from the Earth's surface to avoid its warming is impossible by the laws of physics. "Therefore, the rate of global environmental destruction cannot be reduced by changing one energy source for another while maintaining or increasing the source's power. Improvement of the ecological situation can occur only with a reduction in energy usage power to the ecological limit" (Gorshkov 1995, pp. 402–403).

In this way, according to Gorshkov, the highway out of this ecological dead end runs through the restoration of destroyed ecosystems, forest ones first of all, which naturally does not exclude the introduction of innovative energy technologies, especially for energy conservation. And he envisions this route in terms of a concrete scenario that allows us to stop further accumulation of carbon dioxide in the atmosphere even if we continued burning fossil fuels at year 2000 level volumes, the year this scenario was calculated (Fig. 15.2b).

This scenario requires people to reduce consumption of net primary production from the current 36 to 29%—by 7%. In that event, CO_2 emissions from human-utilized land decrease to 5.4 GtC per year, and total absorption of carbon emitted into the atmosphere (5.4 GtC/year +3.3 GtC formed by fossil fuel combustion not absorbed by the ocean) can be provided by the ocean biota and undisrupted land ecosystems. And then, as you can see from the diagram, further accumulation of atmospheric CO_2 could be prevented.

But it's easy to limit consumption of net primary production on paper. How about in real life? After all, against the backdrop of a growing deficit of land suitable for cultivation, the specter of famine once again stalks the population of many developing countries despite the successes of the green revolution. Therefore, it is clearly harder to reduce the area of cultivated land than to reject the use of raw hydrocarbons. Thus, in this connection, it is more realistic to imagine limiting exploitation of forests used for economic activity, which today makes up half (18 of 36%) of consumed net primary production. Reaching the above mentioned 7% mark requires people to remove 40% of exploited forests from economic use (18% × 0.4~7%), liberating the corresponding portion of settled territory from the human presence and thereby making possible the restoration of forest ecosystems harmed by man's deformation.[2] In particular this concerns forests of the tropical belt, whose productivity is equivalent to four times the unit area occupied by forests and bogs of the temperate belt. With regard to the world timber harvest, some portion could be replaced by artificial materials, and more modern energy sources can be used in place of firewood (Gorshkov et al. 2000, p. 171).

---

[1] The law establishing the physical dependency between a body's temperature change and its irradiance.

[2] Let us note that we are not talking about artificial recultivation, but the natural rebirth of ecosystems on the basis of evolutionary processes. Only such naturally arising biotic communities are able to compensate for environmental disruption in accordance with Le Chatelier's Principle. Artificial agrocenoses such as cultivated forests, as a rule, involve an arbitrary selection of species and are not only incapable of providing for their own stability, but also, as a result of high biological productivity, themselves serve as a source of environmental disruption.

Clearly, in terms of reaching global sustainability, such a measure represents a certain palliative. However, it would allow humanity to give some breathing room to nature in exchange for the time we need to solve other fundamental global problems such as stopping and reversing worldwide population growth, changing the model of consumption in developed countries, universal introduction of energy-conserving technologies, etc.

***

And now let us temporarily distance ourselves from theoretical discussion and try to contrast natural conditions in the different countries of the world from the point of view of their potential transition to sustainable development. After all, while we may consider the World Ocean our common patrimony, inviolate land ecosystems that serve as the primary anthropogenic carbon sink belong to distinct countries where, by the grace of unfolding circumstances, they have managed to be preserved. Here we are aided by the statistical ratings of the global Footprint Network (GFN), an international nongovernmental organization that follows trends in global footprint accounts and bio-capacity status for different countries and regions. In Chap. 3 we introduced GFN data regarding the states that occupy the top niches in lists of "ecological debtors" (whose ecological footprint exceeds bio-capacity) and "ecological creditors (with the opposite ratio). Here we will name them again.

China heads up the first of these groups with an ecological footprint 1652 million global hectares (gha) over its carrying capacity. That's more than twice as high! Next comes the USA (carrying capacity deficit: −1274 million gha), Japan (−532 million gha) and India (−469 gha). Granted, when recalculated per capita, the same rating looks a bit different: Japan - 4.1 gha/person, USA—3.7 gha/person, China −1.2 gha/person, India −0.4 gha/person.

After them comes Germany at −260 million gha (−3.1 gha/person), Italy at −228 million gha (3.9 gha/person), England and South Korea at about −217 million gha (−3.5 and −4.6 gha/person, respectively), and a total of about 100 countries in debt to the biosphere to one extent or another. The countries occupying the first eleven places on that rating list make up 53% of humanity's total ecological footprint.

If we look at that list, we see that most members of this club form part of three **global centers of environmental destabilization** (Fig. 15.3):

- *European*, including the states of Western, Central and Eastern Europe (and excluding Scandinavia), as well as European Russia, Ukraine, Belarus and the Baltic nations—an overall area of 8 million km² with 8% preserved ecosystem.
- *North American* with the USA (minus Alaska), southern Canada and northern Mexico—9 million km² with less than 10% preserved ecosystems.
- *South and East Asian,* bringing in China (except for Tibet), the Indian subcontinent, Japan, the Koreas, Indochina as well as the Philippines—seven million km in all with less than 5% undisturbed ecosystems (Maksakovskiy 2008, Book 1).

It's hard not to notice that of the three centers, two belong to the industrially developed states of Europe and North America, while on the other side of the world,

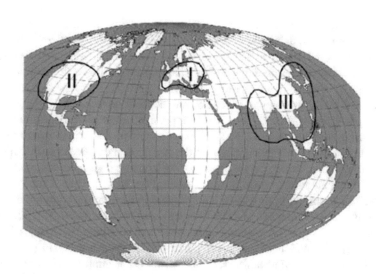

Fig. 15.3 Centers of global environmental destabilization (as described by K. S. Losev): I-European, II-North American, III-South-and-East Asian

we have the developing countries of Asia with high population growth and low-to-medium standards of living. This third region also includes the very different states of Japan, South Korea, Taiwan and Singapore.

The European and Asian centers have roots running deep into history. These were the domains of ancient civilizations, and so the natural environment has undergone great anthropogenic pressure over the long course of changing centuries. Thus the forests of the Apennine Peninsula were wiped out back when Rome stood at its apex. Western and Central Europe followed in the Middle Ages as agriculture quickly developed, cities sprung up, and ironwork demanded charcoal for smelting.

The discovery of America and consequent industrial revolution sharply accelerated the process of ecosystem destruction on both continents, for which the United Kingdom might serve as a textbook example. As it is written in some history books, sheep "devoured" England's forests. Indeed, textile factories, first appearing in the eighteenth century, demanded more and more wool, and shepherds' pastures were created at the expense of clear-cut forests. The forests also went to build the British fleet, as well as metallurgy. Ever since then, the United Kingdom has been a deforested country with the remains of forest masses covering less than 10% of the country, mainly in the northeast part of Scotland.

But while economically successful countries ran through the peak of their nature-destroying activity in the late nineteenth and early twentieth centuries and now have sufficient means to invest in a partial restoration of what has been destroyed, the catch-up states of East Asia have only recently begun to address issues of nature. At the same time, mounting problems connected to overpopulation and the impoverished state of the environment keep these nations in fetters, unable to maneuver.

So, for example, the most acute problem for China, India and the countries between them is the depletion of agricultural land and shortage of fresh water, the vast majority of which (up to 9/10) is expended on irrigation. Water deficits threaten China with particular danger, becoming what might be called a life-or-death question despite coverage of its territory by a lush network of rivers and having the fourth most plentiful water resources of any country on Earth. At the same time, China's per-capita water usage—460 m^3 per person each year—falls within the 90th percentile among the world's countries. Things stand no better with regard to farmland. Tillage shrank by the beginning of the twenty-first century to 0.07 ha/person. This demographic burden on the land has led to tragic consequences, first among them, soil erosion and desertification. According to data from China's Desert Research Institute, these processes have come upon 13 provinces in the country's north and northwest at a rate of 1500 km^2 per year (Maksakovskiy 2008, Book 2).

With regard to preserved nature, China has received a bitter inheritance from its previous regimes. When the People's Republic of China was founded in 1949, forests covered a mere 8–9% of its territory, and the South China rainforests had been almost completely wiped out. The periods of the Great Leap Forward and Cultural Revolution took an enormous toll on the environment. To resolve grain shortages, the central government ordered the plowing of millions of acres of pastureland and the clear-cutting of forests, including in the headlands of the Yangtse and Yellow Rivers. It then comes as no surprise that catastrophic floods have become a common backdrop of modern Chinese life, and yearly losses to natural disasters reach one-fourth of the state budget (Maksakovskiy 2008, Book 2). China has only managed to buck this trend in the past 15–20 years, when, thanks to the adoption of nature conservation laws and reforestation measures, the area occupied by forest (mainly secondary) grew to 14%.

Of special importance to the structure of China's ecological footprint is the carbon portion, 1612 gha. The main polluters are thermal power plants, which run on coal, along with domestic and industrial furnaces. In quantity of carbon dioxide emissions released into the atmosphere, China occupies second place in the world after the USA. At the same time, it is the greatest worldwide emitter of another powerful greenhouse gas, methane, which finds it source in coal mines and several branches of agriculture, especially rice-growing and livestock raising.

Many ecological problems unite China and India, which occupies second place in population and third in ecological footprint (1063 gha). First of all, these common problems include degradation of agricultural land and insufficient land resources, due to which tillage adds up to a mere 0.17 hectares per person. Granted, the water issue does not affect India as acutely, but it will grow more pressing with time. While fresh water resources are judged sufficient for now, a growing population will demand a constant increase in water diversion to fill its needs, for irrigation, first of all. Along with this, there is a high level of pollution in surface waters from industrial and domestic runoff, which as a rule flows into rivers without purification, and water is being removed from aquifers at twice the rate of replenishment. As a result, water tables are falling by 1–3 m/year, and, as specialists estimate, this house of cards may collapse at any moment. After that, grain production in India will fall by more than a quarter (Danilov-Danil'yan and Losev 2006, p. 138).

Table 15.1 "Ecological debtor" countries of South and East Asia. The order of the countries corresponds to their biocapacity deficit

Country	Ecological footprint millions ha	Bio-capacity, millions ha	Biocapacity deficit, millions gha/per capita		Population, millions	Population density, people/km²	Population growth rate
China	2959	1307	−1652	−1.2	1336	141	0.5
Japan	602	76	−532	−4.1	127	334	−0.3
India	1063	594	−469	−0.4	1164	362	1.3
S. Korea	233	16	−217	−4.6	47	489	0.2
Thailand	158	77	−81	−1.2	67	130	0.6
Malaysia	129	69	−60	−2.3	26	87	1.6
Philippines	115	55	−60	−2.3	88	339	1.9
Pakistan	132	74	−58	−0.3	173	253	1.6
Bangladesh	98	59	−39	−0.25	157	1101	1.6

Source: The Ecological Footprint Atlas 2010, Worldstat info. Data on Population, ecological footprint and bio-capacity based on 2007 figures, numbers rounded

Unlike China, India has 20% forest cover, but this fifth of the country's territory is largely secondary forest, scrubland and man-made savannah. At that, it should come as no surprise that, under conditions of land shortage, deforestation reaches a rate of 1.5 million ha/year. The reason behind this lies at the surface. It is a critical need for tilled land and the use of timber as fuel. At the same time, the country's demand for firewood and industrial lumber stands seven times higher than what could be provided by the natural growth of forest resources. Thus, the cutting down of forests will clearly continue in India, with the elimination of virgin forest in the Himalayan foothills arousing particular worry (Maksakovskiy 2008, Book 2).

In Table 15.1, we have also presented a number of other countries in the region that have the greatest deficits in biological capacity and heaviest ecological footprints. The group is extremely diverse. Along with impoverished countries primarily made up of subsistence farmers practicing semi-natural agriculture, such as Pakistan, Bangladesh and the Philippines, it also includes economically developed Japan with South Korea right behind it, members of the "Asian tigers"—states that made a decisive breakthrough in development in the later twentieth century. There are countries such as Malaysia, where rich forest vegetation still covers half of the territory, but the same countries also belong to the number of world leaders in forest destruction (Malaysia, Thailand, the Philippines). As a percentage, this crown goes to Bangladesh, which destroys 4.1% of its forest land each year, while occupying second place in absolute rate of deforestation. Pakistan and Thailand follow at 3.5% each year, and then the Philippines at 3.4% (Maksakovskiy 2008, Book 2).

Another commonality of the countries figuring in this list is high population density, which more often than not surpasses that of Western European States. Bangladesh also claims a tragic leadership here occupying one of the highest positions in the world by this indicator. And if you combine this with a high birth rate (Philippines—1.9%,

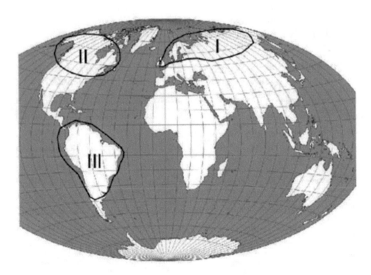

Fig. 15.4 Centers of environmental stabilization (as described by K. S. Losev). I-North Eurasian, II-North American, III-South American

Pakistan, Bangladesh, Malaysia—1.6%), you come to understand the myriad of pitfalls this region must step across in its path to sustainable development, beyond accounting for its role as the weightiest climate destabilizer on the planet by scale of CO_2 emissions (Zalikhanov et al. 2006).

And, now, let us cast our gaze upon the other end of the scale, on those countries belonging to the number of "ecological donors," where ecological capacity exceeds ecological footprint. Beyond comparison in biological capacity size stand Brazil at 1708 million gha, Russia at 816 million gha and Canada at 492 gha, occupying the top three places in the world rating for this indicator (See Chap. 3, Table 3.1), and by no coincidence. After all, it is their territories where you find the world's largest untouched masses of forest, with an overall area of more than 16 million km². And when combined with the forest ecosystems of their surrounding regions— Scandinavia, Alaska, and Amazonian Bolivia, Colombia, and Peru—we see the contours of the three **global centers of environmental stabilization** (Fig. 15.4):

- North Eurasian (11 million km²), which includes the north of Scandinavia and European Russia as well as most of Siberia and the Russian Far East except for its southernmost part;
- North American (9 million km²), including Northern Canada and Alaska;
- South American (10 million km²), Amazonia and the neighboring mountain territories (Maksakovskiy 2008, book 1).

It is these, along with the World Ocean with its still little-disturbed ecosystems, that, like the turtles ancients once imagined to carry the Earth upon their backs, play the decisive part in maintaining the stability of the biosphere, allowing it to more or less successfully resist the yearly mounting anthropogenic pressure.

If we go down the list of countries that, along with the three above-mentioned, carry the most weight in stabilizing the Earth's environment, then we see that most of them also belong to the number of countries richest in forest resources. Just ten countries of the Southern Forest Zone possess three-fourths of the world's tropical rainforest. These would be Papua New Guinea (63% forest cover), the Democratic Republic of the Congo (57%), Brazil (55%), Peru, Colombia, Bolivia (52% each), Venezuela (52%), Myanmar (46%), Indonesia (44%), and India (23%). Granted, that last one, with its large ecological footprint, belongs to the biospheric "debtor" column.

There is, however, a paradox in the fact that many of the countries in this top ten are also leaders in yearly removal of forests. First among them stands Brazil, at an astonishing 20,000 km^2 of forest each year. This list of "record holders" includes Indonesia (10.8 thousand km^2), Bolivia (5.8 thousand km^2), Venezuela (5000 km^2), Myanmar (13.4 thousand km^2) and Paraguay (3.3 thousand km^2) (Maksakovskiy 2008, Book 1). In all, according to data from the FAO, for the period from 2000 to 2010, the world lost 3.2% of all its year 2000 forests, an area as large as the Republic of South Africa. When accounting for forest restoration, net losses came to 1.3% (State of the World's Forests 2012 pp. 19–20). And while a recent FAO overview on forest resources, presented by its general director at the 14th World Forestry Congress in Durban, South Africa, in 2015, expressed hope for positive changes, this is the type of cautious optimism one acquires when wishing to take the desired result for the true one.

Granted, on the plus side, the trend toward deforestation in roughly half of the world's countries has been stopped or even reversed. As a result, net global forest losses have fallen from 0.18% in the early '90s to 0.08% in 2010–2015, i.e. by more than 50%. And global carbon emissions linked to forest degradation fell by 25% over a decade and a half—from 3.9 GtC per year in 2001 to 2.9 GtC per year in 2015. Brazil achieved the most impressive successes in this area, accounting for over half of the reduction in carbon emissions (Global Forest Resources Assessment 2015). But, nonetheless, according to the latest satellite data, tropical forests continue to disappear at an accelerated rate, falling by an additional two thousand km^2 each year (Hansen et al. 2013). On the whole, the tropical belt carries about half of the world's total forest loss, and it is here that we observe the most staggering ratio of losses to growth (see Table 15.2). We witness the highest rate of forest degradation in the countries of South America, especially Argentina, Paraguay and Bolivia.

As you can see from the table, the overall area of forest losses over the 12 years added up to 2.2 million km^2, while growth reached just 0.8 million km^2. Tropical forests disappear or are degraded fastest of all. Causes for forest loss differ materially depending on climate zone. While forest fires cause most of the losses at moderate and northernmost latitudes, human activity occupies the primary role in the death of tropical forests, especially intensive clear-cutting meant to free up land for agricultural needs or obtain timber for fuel and commercial use. For more than two billion people on Earth, firewood serves as the main source of household fuel. In Africa, 80% of cut timber goes to this purpose (FAO 2010).

Unfortunately, the short term gain to be had from cutting down forests, measured in current market prices, pales in comparison to the long-term and less obvious advantages of sustainable forestry. Many of these advantages, such as climate

Table 15.2 Overall number balance of forest degradation and growth by climate zones, 2002–2012 (Hansen et al. 2013)

Forest by zone	Losses (km²)	Increase (km²)
Tropical	1,105,786	247,233
Boreal	606,841	207,100
Subtropical	305,835	194,103
Temperate	273,390	155,989
Total	2,291,851	804,425

Source: Naymark (2013) and Hansen et al. (2013)

stability, water system maintenance, defense against soil erosion, or preserving biodiversity, have no market price at all, and as such are not compensated by corresponding economic mechanisms. As the UNEP analytical group stated in their report, *The Economics of Ecosystems and Biodiversity*, "In most countries, these market signals do not take account of the full value of ecosystem services; moreover, some of them unintentionally have negative side effects on natural capital" (TEEB 2010, p. 27). As a result, cost assessments of forests often go down, creating the preconditions for their unencumbered elimination.

Overall, forest usage is an area where national and human interests deeply conflict with those of private business. Here the safeguard must, primarily, be provided by the state. Brazil's experience testifies to this by a certain measure. All the way back in 1988, that country developed its forest defense program, including the repeal of subsidies and credit for agricultural projects in the Amazon Basin. In 1996, Brazil's legislature adopted a special law meant to hold back the process of tropical forest destruction and forbidding farmers to remove more than 20% of virgin forest growing on their land. Thanks to these efforts, Brazil has managed to slow the rate of forest extirpation by half since 2005—from 40 to 20 thousand km² per year (Maksakovskiy 2008, Book 2; Naymark 2013).

Another cause for mass perishing of forests in recent decades has been epidemics spread along with tree parasites. In the opinion of British specialists, this comes as a side-effect of opening international trade and transport routes. As a result, the number of introduced species constantly increases, including those of a pathogenic nature to which local plants have not acquired immunity. Thereafter, illnesses in trees often grow to epidemic proportions (Boyd et al. 2013).

Earlier, we introduced calculations by Viktor Gorshkov concerning the possible end to the process of accumulating atmospheric CO_2 in the event that 40% of exploited forests are removed from economic use. Unfortunately, far too few countries have the corresponding resources to achieve this purpose. And so, in our view, we must put our hopes in the countries of the northern and southern forest belts. Their broad expanses of wooded territory and areas of low population density will allow them, at limited cost to themselves, to emancipate a part of their destroyed forest ecosystems form anthropogenic pressure. Among others, these countries include Russia, Canada, most states of South America, and the Democratic Republic of the Congo, each capable of making a distinct guarantee for global environmental stability.

The great misfortune, however, is that for most tropical countries, the primary task concerns less the increase of their forest wealth and more its defense from predatory destruction. And if developed countries are seriously troubled by the dangerous degradation of the biosphere, they cannot stand aside as this problem unfolds. All the more so because, as powerful sources of greenhouse gas emissions, they are practically living off a kind of "ecological rent," extracted from countries with preserved ecosystems, i.e. off of the exploitation of other people's ecological resources.

Unfortunately, this problem has still not been recognized on an intergovernmental level, just as, up to now, there is still no complete transparency regarding the true price of what we call ecological resources. In any case, it is not the price that city-dwellers pay, for example, for water (i.e. to collect and supply the water plus amortization of facilities), or what farmers and real-estate owners pay for rent and mortgage. After all, the latter's price is not determined by the natural communities that "take up house" there, so much as supply and demand for the concrete "services" the land is able to provide. That might be, of course, a picturesque landscape of some wooded area where a resort could be constructed, but more often we are speaking of a plot of land's fertility (when accounting for agricultural suitability), location (usefulness for urban or industrial construction), or mineral wealth. In this way, assessments almost always concern some financial advantage or another that might be obtained from a developed plot of land and not, as a rule, its overall biological value as an element of the biosphere, fulfilling its share of the work to maintain stability.

In the late Soviet Union, at the end of the 1980s, an attempt was made according to expert assessment to determine the resource-conserving and resource-restoring role of national parks in monetary terms, i.e. to evaluate how they contributed to environmental stabilization. In the experts' opinion, the land could be assessed at 2000 rubles per hectare, or $1000 per hectare when recalculated for the established (non-market) course of the US dollar (Losyev et al. 1993). In this way, they valued the work carried out by ecosystems—and national parks are textbook examples of undisturbed ecosystems—their function in maintaining environmental stability, very conditionally, to be sure, at $1000 per hectare.

Meanwhile, for 2014 alone, on the 86 million km^2 of partially and completely destroyed ecosystems, humanity obtained a gross product on the order of $78 trillion (CIA World Factbook n.d.). Therefore, one hectare of deformed or destroyed ecosystems contributes to the gross world product at a rate of about $9000 per year. Accounting for inflation, which by statistical estimates has been 50% over the past 25 years (Statbureau.org 2007–2015), we could equate this to $4500 in 1988–89 prices. The difference, we see, is 4.5 times. Even when accounting for inaccuracies of comparing market and non-market values for the dollar, it is still very real. And so any economist unconcerned with the environment can, if they wish, easily "prove" the disadvantageousness of conserving natural ecosystems.

But this is today's disadvantage, and that line of thinking now arouses doubt among many people of common sense even without a particular knowledge of ecology. After all, clean river water, forests of berries and mushrooms, and the joy of experiencing unencumbered wild nature are all disappearing from the world of modern humans. A future-focused life, with thought for grandchildren and great-grandchildren, obviously demands a different strategy. And since hope for ecological stability on the territory of one's own country, "fenced off" from the rest of the world as it exists, a number of developed countries' plans for sustainable development cannot withstand serious scientific argumentation, their populations will sooner or later have to make a decision. Better sooner than later. A decision between a boom of unrestrained consumption and reasonable self-limitation. Between expensive and ambitious projects such as corporate skyscrapers, fashionable flagship hotels or spectacles observed by millions and investment in the business of protecting the environment which promises no quick profit but which should, in theory, be a point of prestige for big and medium companies. That is to say it is a choice between corporate and national egotism on the one hand and concern for the fate of humanity on the other.

And so all of this doesn't look like a tangential abstraction, let us try to make it concrete with the example of Russia, whose geographic position and natural particularities provide it a key role in the global ecological layout.

***

Like any industrialized country, Russia makes its own contribution to the pollution and degradation of the environment, including anthropogenic carbon emissions, especially those from fossil fuels. Fossil fuel combustion makes up about 80% of Russia's greenhouse gas pollution. In 2000–2001, these emissions were estimated on the order of 0.54 GtC/year, 5.7% of the world total. At the same time, net absorption of atmospheric CO_2 by Russian forests added up, by different estimates, to anywhere from 0.2 to 0.5 GtC/year. The summary flow of CO_2 to all the country's undisturbed ecosystems added up to 1.0 GtC/year. From this, it follows that not only does Russia not serve as a source of increase to CO_2 concentrations in the atmosphere, it provides an unused resource of its absorption, of a size estimated to be on an order no less than 0.3 GtC/year (Contribution of Russian Forests..., 2004; Zalikhanov et al. 2006). Obviously, other states are using this resource, whose available area of preserved ecosystems is not adequate to the task of absorbing their own CO_2 emissions. In other words, Russia's ecological space is offering them a free ride.

Yet rather than solidifying its natural position and status as an ecological great power, Russia has moved ever more distant from its past environmental priorities, which, in the early 1990's allowed it to join the worldwide movement toward sustainable development. Domestic policy in the years from 2000 to 2015 we cannot call otherwise than de-greening. Meanwhile, cause for concern with the ecological situation in the country has not, in any case, diminished.

So, for example, despite 11 million km^2 of territory with preserved ecosystems (about 65% of the country's whole area and 22% of territory with undisturbed ecosystems worldwide) two-thirds of residents live in environmentally degraded areas, home to hundreds of cities including the nation's largest. In most of them are found permanently unacceptably high concentrations of toxic substances in both the air and drinking water supply sources. But while, in the 1990s, pollutant emissions into the atmosphere gradually decreased, from the year 2000 they began growing yearly. And in urban and suburban garbage dumps and areas little different from the typical garbage dumps, 110 billion metric tons of solid industrial and domestic waste have accumulated, poisoning the groundwater as well as surface water sources and filling the air with dangerously unhealthy dioxins (Danilov-Danil'yan 2006). You wouldn't see such things in any other developed country in the world today. Thus a degraded ecological backdrop has become for the majority of Russia's population an everyday reality.

This reality, however, does not join itself in the average Russian's consciousness to the issue of sustainable development, having no relation, so he imagines, to his life today or tomorrow. Is that not because the Russian citizen's perception of tomorrow, as research by the Club of Rome has shown, does not usually extend further than a few weeks? To go along with this, both television and the press, in tune with mass psychology and, to no small extent, forming it, fill up their channels and pages primarily with so-called breaking news, practically starving all thought of the country's fate, the future of its people or of the world as a whole.

As concerns statesmen and politicians, the majority of them suppose that, in a period of structural reform, questions of sustainable development are not as relevant to Russia. The logic here is simple: first we need to overcome the difficulties of today, and then think about a transition to sustainability. Furthermore, with the apathy of the voting public toward any general ecological problem not directly affecting their town or subdivision, such issues are pushed off the electoral platform. Thus, for a person entering power, it is clearly not a winning theme. After an outpouring of environmental activity in the years of perestroika and post-perestroika, interest in ecological problems began to fall sharply from the first or second popular priority in 1989 to the ninth or tenth in 2000 (Losev 2001). And while, in most developed countries, issues of sustainable development consistently fall within the field of public attention, in Russia they remain the narrow province of professionals, only to be discussed at specially dedicated conferences and symposiums.

Highly indicative in this regard are also the metamorphoses that have occurred over the last two decades in Russia's state environmental institutions. While in the first post-perestroika years, the primary institution held ministry status, it was then downgraded to the level of State Committee for Environmental Protection and Natural Resources. After Vladimir Putin was elected in 2000, the Committee, too, was dissolved, and its functions were not given to the Natural Resource Ministry or any other executive organ. In the President's order to dissolve the committee, no successor institution was determined at all, an unprecedented event in the history of Russian bureaucracy. Only after several years did a new article appear in the Natural Resource Ministry's charter, delegating it environmental protection functions. That is, a return took place to the vicious Soviet-era practice: let he who destroys nature regulate it himself.

Results did not delay in coming. Strictness declined in ecological demands of businesses. Access to natural resources was materially eased for large- and medium-sized companies. The best natural nooks around big cities began to be refitted as elite suburbs, and national parks had to constantly defend themselves from encroachments on their seemingly inviolate territory by local governments and their business associates.

In 1994, Boris-Yeltsin issued the presidential order "On the State Development Strategy of the Russian Federation in Environmental Protection and Providing Sustainable Development." Two years later, another presidential order was confirmed, "Concept of the Russian Federation's Transition to Sustainable Development." By that order, the federal administration was directed to design "A Strategy of the RF's Transition to Sustainable Development."

This document, however, foundational for so many counties and recommended for passing by the Rio Conference of 1992, designed in 1997 by the Economic Development Ministry with the involvement of specialists and public figures, did not go on to become law. It was reviewed by the government in late 1997 and sent for final corrections according to standard practice. But then the economic crisis of 1998 interfered, followed by a default in August. The Strategy project was simply forgotten, removed from the agenda. In its own way, though, Russia's failure to adopt the program for transition to sustainable development was symptomatic. What would come in its place?

For today, Russia has a few palliative documents which make a bare attempt to resolve the tasks put forward in the strategy. This included the 2002 adoption of "The Ecological Doctrine of Russia," which then Prime Minister Mikhail Kasyanov presented to the 2002 Johannesburg Summit as some great achievement, though it brought no noticeable results. This was later supplemented by the administration's approval of the "Socio-Economic Development Strategy of the RF for the Period up to 2010" and the parallel adoption of select, more focused development strategies up to 2020. But these documents could not unfortunately replace the country's mothballed strategy for transition to sustainable development.

In 2008 the Ministry of Natural Resources added the words "and ecology" to its name, though ecology itself gained nothing from this changing of the shingle. Today, responsibility for ecological problems is shared between about twenty ministries and federal agencies. But as they say, too many cooks spoil the broth, and the lack of a single organ working on environmental questions and nothing else has told very negatively upon green policy (Danilov-Danil'yan 2001; Fomin 2005). In that respect, Russia is falling dangerously behind the majority of other states. The country's anti-ecological drift has ended in an increase of relative power for nature-exploiting and polluting industries, the growth of energy consumption per production unit, and ultimately the orientation of the economy on an unlimited resource base. But if the economic system carries "anti-ecology" in its genes, then how can state environmental protection resist it while following in its wake?

Especially dangerous in view of this, we witness the assault on Russia's "ecological bastions"—Siberian taiga forests and bogs, experiencing ever greater pressure from business. As stated in the *Atlas of Russia's Intact Forest Landscapes* (Aksenov et al. 2003), one often happens to hear that Russia is a country where the greater part is still covered in endless slumbering forests, practically unpeopled and untouched. That is what many experts in the area of environmental protection think when they say that two-thirds of Russia's whole forest zone is made up of "completely wild nature."

Sadly, the atlas concludes, "These findings refute the myth that ancient or virgin forests still dominate Russia. Such forests now dominate only the northern parts of Eastern Siberia and the Russian Far East...In most parts of European Russia and Western Siberia, and the southern parts of Eastern Siberia and the Russian Far East, the forest vegetation has been fundamentally transformed by human activity. No large intact landscapes remain in many of these western and southern areas, while the intact forests that remain are broken up into fragments, too small to sustain the full array of components and functions characteristic of a natural forest landscape" (Danilov-Danil'yan 2001; Fomin 2005).

In Russia, however, not only is little thought given to maintaining the sparsely-populated regions of Siberia and the Far North, but just the opposite. Preparations are made for their further conquest. One such project, the planned construction of the Evenkiskaya Hydroelectric Station on the Lower Tunguska, to be one of the world's largest dams with an output of about 8 gigawatts, a height of 200 m and a flood area on the order of a million ha. Construction time is estimated at 18 years. Planned simultaneously are two high-voltage power lines running from the dam to Tyumen, at distances of 600 and 800 km, as well as the installation of three additional generators at the Nizhe-Kureyskaya Hydroelectric Station to provide energy requirements for construction (Nefyodov 2008). And that is on territory recognized by UNESCO as the environmentally cleanest on the planet!

Mikhail Lomonosov once uttered the prescient words that, "Russian strength will grow through Siberia." But today it looks like we are undermining this strength, understanding it too narrowly and literally. Although it would seem the situation of ecological crisis itself is giving the old prophesy a new meaning and life. After all, through the nature of Siberia and the Far East, preserved inviolate, Russia will carry its weight in preserving the whole biosphere, stabilizing concentrations of atmospheric CO_2, maintaining the continental water cycle, providing soil formation and so on. Thus without dredging up any more oil from the Earth's depths, which we have freely squandered not thinking of our descendants, and without any more nuclear weapons, but precisely with this priceless patrimony, can we affirm our status as a world power, and at the same time, like the many countries that have lost their natural resources, look concernedly into the future and try to dodge the threat approaching from further environmental degradation.

By the way, one should not understand the word "priceless" to mean that it has no concrete price at all. Experts have tried to calculate, measure and weigh out this price. They base them on the economic losses incurred by the US, Japan and the countries of Europe in the course of fulfilling the obligations of the Kyoto Protocol for lowering carbon emissions. These expenses, as the international specialists determined, add up to between 550 and 1100 dollars per metric ton of non-emitted carbon. And if, as we said above, the summary flow of Russia's anthropogenic carbon into its own ecosystems equals about 1 GtC/year, surpassing CO_2 emissions on Russian territory by 0.3 GtC/year, then Russia's ecosystems also remove other people's carbon from the atmosphere at a value of between 165 and 330 billion dollars a year. These are billions that our country is practically investing in the world community, including in developed countries (Zalikhanov et al. 2006).

Carbon sources and sinks were evaluated simultaneously for other continental-scale regions as well—Asia (besides the CIS), former Soviet Republics, Africa, Western and Central Europe, the Americas and Australia (Kondrat'yev et al. 2005). As has already been said, today South and South-East Asia serve as the greatest destabilizers of the Earth's climate. North America and Europe follow behind them, largely due to industrial emissions. Africa stands close to a neutral result for the time being. Australia and especially South America remain areas that stabilize the climate thanks to the undisturbed ecosystems preserved there. The balance of anthropogenic carbon within the CIS is close to even due to Russian ecosystems.

In Table 15.3, we have shown evaluations anthropogenic carbon sinks in undisturbed and partially disturbed forest ecosystems of the Northern Hemisphere (Russia, USA, Canada), as determined by various methods and contrasted with satellite data (Myneni et al. 2001).

Table 15.3   Carbon sinks in forests of Canada, Russia and the USA

	By satellite date			By other data		
Country	Stores, GtC	Sinks, GtC/ year	Area, ha (millions)	Stores, GtC	Sinks, GtC/ year	Area, ha (millions)
Canada	10.56	0.0731	239.5	11.89	0.093 0.085	244.6
Russia	24.39	0.2836	642.2	32.86	0.429 0.058	816.5 763.5
USA	12.48	0.1415	215.5	13.85	0.167 0.098 0.020	217.3 247.0

Source: Zalikhanov et al. (2006)

Unfortunately, the focus of the Kyoto Protocol as well as the 2015 Paris Accord that replaced it on CO_2 emissions from fossil fuel combustion turns to the disadvantages of a number of countries, including Russia, through the clear undervaluation of factors such as carbon emissions that result from land usage or their deposits in preserved ecosystems. In the meantime, the interests of stabilizing the global environment demand a fundamental revision of our attitude toward planetary wealth. First of all, this means more efficacious measures for investing financial capital in the protection and restoration of forests which would outbalance the advantages of annihilating them. Or as Viktor Gorshkov proposes, introducing an international tax that developed countries would pay to countries possessed of a biota untouched by civilization on a scale that exceeds the potential profit of exploiting it (Gorshkov 1995, p. 36). And then, perhaps, in Russia a different scheme of motivations would go to work, and the damage russians are doing to our national interests would become obvious. Preserving the forest masses of Siberia, the Far East and the north of European Russia would become a strategic priority.

If you compare the area taken up by undisrupted and little-disrupted ecosystems in Russia at the start of the twentieth century, before the first symptoms of the ecological crisis, with today, you get the ratio of 80–65% (Losev 2001). Therefore, over the century and mainly over the course of the 70-year reign of centralized economics, the country lost 15%, or 2.5 million km^2 of its ecosystems. In this a significant role was played by an irresponsible attitude toward the land, which was generously given to any project, needed or unjustifiable.

You can still recognize the results of this "management" today. The area of agricultural land per capita, 1.5 ha, of it 0.88 ha under the plow, spreads twice or three times as wide as in most developed countries. In Finland, for example, the same indicators stand at 0.44 and 0.43 ha, Sweden—0.35 and 0.29 ha (Worldstat Info n.d.), and these northern countries are crop exporters. While the rest of the world experienced a Green Revolution with its intensive farming methods and reduced area of tilled land, Soviet agriculture went down the opposite path, expanding tillage at a low level of production efficiency.

In his tract, *Rebuilding Russia* (Solzenitsyn 1990), Aleksandr Solzhenitsyn brings in the words of famed early twentieth century statesman Sergey Kryzhanovsky, a supporter of Prime Minister Pyotr Stolypin and final State Secretary of the Russian

Empire: "Native Russia is not disposed of the store of cultural and moral strengths for the assimilation of all outlying territories. It is sapping the Russian national core." No, at the moment that essay appeared, the ecological aspect did not concern its author, having written of "the spiritual and bodily salvation of our people" appearing still before the fall of the Soviet Union. But if instead of the nation's outlying imperial territories—the former Soviet Republics—we put the sparsely populated regions of Siberia, the Far East and northern European Russia, the ideological vector of Solzhenitsyn's musings in many ways joins up to that which troubles today's ecologists. Imperial ambitions weigh ever more upon our national consciousness, including our relationship with nature. A gradual retreat, a leaving behind of these regions like never before coincides with the task of propping up the well peopled Russian lands occupying the space within the so-called "triangle of cities"—St. Petersburg, Irkutsk and Sochi. And if we add to that Vladivostok and the greater part of Primorsky and Khabarovsk Krais, we get the areola of ideal natural and climatic conditions in which 95% of the country's population lives.

On this territory of 5 million km², stretching south from the 55th or 60th parallel and including the Urals, European Russia and the southern parts of Siberia and the Far East, is concentrated 95% of the country's industrial potential and 100% of its agricultural wealth. For the great majority of the population, that makes the process of liberating weakly conquered territory practically painless (Losev 2001). Along with this, the native peoples of the North will benefit, being today for all purposes on the edge of extinction or faced with the choice of leaving their homelands and fully losing their cultural identify. First of all, they will regain, at their disposal, a pure natural environment, and, secondly, the possibility to return to their cultural roots and traditional ways of life.

With regard to the masses of wild nature left whole, they should all, just like national parks, be restored to state property with a total ban on their economic use beyond special cases—the development of geological deposits with low-impact technologies, construction of strategically important objects, etc. But compensatory mechanisms should be arranged for them in the form of ecosystem restoration on another, naturally similar territory. A ban on economic activity within undisturbed ecosystems does not mean depriving people of the opportunity to communicate with wild nature, so long as they strictly obey certain rules (ecological tourism), the most important of which is performing any type of activity on reserved territory exclusively under their own kinetic power.

***

All of this, however, is not the end of Russia's ecological potential which it could direct to the welfare of the rest of the world. As Russian geographer Boris Rodoman notes, the landscape of the average Russian *glubinka,* or deep backwater, fundamentally differs from the countryside of most European countries. The fact is, centuries of a strict power vertical on the territory now called post-Soviet Space, sparsely peopled "dead zones" arose around the edges of most provinces, where traditional settlements disappeared in the Soviet era. As a result, a unique "polarized" landscape formed. As power flowed "vertically" from Moscow down to the provincial administration centers, the "horizontal" infrastructure weakened decidedly

(Rodoman 2012). For good reason does one hardly encounter the understanding of a glubinka outside the Russian language.

In the ancient Roman Empire, they said "all roads lead to Rome." A similar situation arose, too, in the Soviet Empire, and in post-Soviet Russia, which is particularly clear from non-Chernozem areas. So while half a century ago, every village was linked to its neighbors by three or four roads serviced by motorized transport, today you can only get to them on foot, or by bicycle at best. In more-or-less tolerable condition are only those roads along which the leadership travels. Along this "vertical axis", too, run the bus routes, oriented in the direction of the nearest provincial or district center. The closer you get to the district's edge, the quieter and more transparent life becomes, thus at direct proximity to the edge, especially at the crossroads of several provinces, do you see the formation of true backwoods practically without any transport network. In its place forms a natural network of transcontinental ecological corridors, a kind of econet uniting the territories of undisturbed and little-disturbed ecosystems.

For good reason do we use the term transcontinental, since that is precisely where the econet aims. Rodoman writes, "To support the vitality and wholeness of the biosphere, natural acreage should occupy not only enough area but make up a solid, contiguous mass by way of green corridors" (Rodoman 2012). But while in Western European countries, the realization of such an idea would require purchasing and recultivating land in private holding, we have no need to exert such efforts. It is enough to maintain the status quo, that is, not to restore deserted roads but to keep what exists in order, providing transport along axial highways. And there is no surer way to kill off nature than to build a road through it.

Rodoman points to one more reservoir of barely-touched wild nature. That would be military bases and training grounds. He writes, "The Russian Ministry of Defense is the world's biggest consolidated landholder. Inside the barbed wire of our forbidden zones, judging, for example, by greater Moscow, could lie a whole tenth part of our country." At that, the officers are even better preserving the natural landscape they occupy, so long, of course, as they don't use it for its intended purpose, than the impoverished and scarcely dependable National Park Service. Therefore, along with state-run dachas, hunting reserves, oligarchs' estates, and so on, it is these military training grounds that de-facto serve as our truly protected nature preserves, and it is in our interests that these lands remain under the defense and security organs as long as possible.[3]

But of particular interest in Boris Rodoman's concept comes its geopolitical aspect, despite a number of controversial points contained within. We are speaking of Russia's ecological specialization, as the cited article is titled as well. Though the ideas he expresses are not entirely new and have already come under discussion by Russian ecologists (Losyev et al. 1993; Kondratyev et al. 1993; Arsky et al. 1997; Danilov-

---

[3] In fairness, it is worth recalling that a famous ecologist wrote about this in the late 1980s, during perestroika, correspondent member of the Russian Academy of Sciences Alexey Yablokov. Granted, then the question was of giving over the territory of defense objects to the Russia environmental protection system, which, unfortunately, met with little understanding from either government bureaucrats or officers.

Danil'yan 1999), the topic has received little hearing since the year 2000. Under the conditions of a de-ecologizing Russian politics, many participants of that discussion had to switch to questions of the environmental protection system's survival. The role Russia could play in preserving the planet's ecological balance went onto the back burner, disappearing from the public eye. Thus the very fact that Rodoman addressed this problem in a new historical go-around serves as proof that the idea itself is still alive, and, contrary to accepted policy, is being reborn in popular consciousness.

In short, Rodoman formulated the dilemma before our country as this: shall we remain in the role of a backward outsider playing catch up, aiming to enter the world system as an equal player at any cost? Or shall we, making use of our geographical advantages, secure a role as the leading ecological donor. "There's no need to try and catch a leaving train, or have to come even with other countries by some economic indicator...Russia could specialize in the role of ecological guardian, protecting the natural landscape in the interests of all humanity. Perhaps we might depend on military force as well, most likely international, to prevent, say, the settlement of our Siberia. We cannot accept to happen there what occurred with Manchuria over a mere century. It was just as much taiga, just as sparsely populated as Siberia" (Rodoman 2004).

Of course, this is not the first time someone has expressed the idea that Russia, with its wide open spaces and relatively low population density, will have a hard time holding onto Siberia and the Far East. But, in the writer's opinion, the situation is not all that simple and, under certain circumstances, could be turned to Russia's advantage—if it can put its rich natural potential in the service of the rest of humanity. To that end, he expressed doubt concerning our established approach toward the issue of depopulation. Does depopulation truly represent an absolute evil to our country, threatening to blot out its future? And does it need a stream of immigrants to supplement lost population and labor power?

Granted, immigration policy could solve a few problems—stabilizing population numbers, providing workers to the fields where they are most needed—but it does not promise fundamental change to the state of affairs in the country. For the simple reason that 80% of its residents are "economically redundant." Rodoman writes that they are unattached to the oil-and-gas pipeline and of questionable potential as producers and consumers, and so destined for degradation and extinction. Whole socioprofessional castes, according to research by economics professor Natalya Rimashevskaya, are sinking like rocks (Rimashevskaya 2003). Thus, against the background of the country's current export-import orientation, we should more likely be speaking of its economic overpopulation. Overall, to see only the negative in Russia's low population means being governed by the logic of bygone days. To fulfill its ecological mission, Russia has a more than adequate population.

Rodoman, to this end, brings in the analogy between individual professional orientation and the specialization of a country or region. Just as each person most reasonably seeks to apply their strengths to the sphere of activity for which they have a calling, so, too, it makes little sense for backward countries to attempt copying the achievements of the economic leaders that have surpassed them, and they should rather realize themselves in the area of least competition, i.e. occupy their own, inimitable niche. For Russia, this niche unquestionably lies in its natural resources, thanks to which it might become the ecological magnet of the whole eastern hemisphere. And here we will use another simile brought forward by Rodoman, comparing Russia to the forest park of a large city. (Today one might look upon the whole Earth's landmass as a type of worldwide city—*Ecumenopolis*.)

"By rejecting 'heavy industry' and 'medium engineering,' our country could march decisively into post industrial society—not into the business center, of course, and not onto skid row, but to the peripheral green zone…Russia could occupy, in relation to Western Europe, the role that park regions around Moscow play for the capital, i.e. take upon itself the global function of 'the world's garden district'—to be a source and reservoir for clean water and air, a place of physical and spiritual recovery for its visitors…"(Rodoman 2012).

Therefore, only in connection with developed countries could Russia obtain a truly worthy position in the capacity of an equal ecological partner. But for that both sides need to recognize the great benefit that such strategic cooperation lays before them. First of all, of course, this means preserving Siberian and Far East taiga as well as other remaining areas of virgin Russian nature as an ecological resource of global significance which often concerns economically advanced countries more than Russians themselves. There are also broad possibilities for ecological tourism, exposure to the world of untouched nature, demand for which grows with each passing year. Rodoman considers, "precisely a global, international approach to the Russia's environmental protection mission would best enable its preservation as a unique country with a very specific civilization" (Rodoman 2012).

Russia, for its part, might stake a claim for corresponding compensation ("a buyout") for declining to exploit its forest resources, apply agrochemicals, pollute the environment with industrial technology or engage in any other ecophobic activity. In other words…for doing nothing. Yes, the eternal Russian question should be reformulated: not as "What is to be done?" but "What is not to be done?" as the country exits onto the path of ecological specialization. And though the mania for that activity—a disease of modern humanity fraught with destructive consequences for the biosphere—has gripped many nations, Russia, it seems, takes home the gold here. Indeed, a rejection of schemes possible and impossible would surely turn to the benefit of our country. Let us merely recall the upturning of virgin lands, the draining out of bogs (which later had to be refilled), the planned redirection of Siberian rivers or other grandiose projects to vanquish nature.

Thus, in the case of a directed ecologization of Russia, it has something to relinquish in both its present and its past. That includes artificially inflated defense spending (the second most numerous army in the world and one of the highest shares of GDP—5.4%—devoted to defense), and the projected construction of the gigantic Evenkiskaya Hydroelectric Station in the wilds of the Lower Tunguska, and the development of new geological deposits which brings harm to the environment (for the sake of carrying out one or two suitcases of diamonds from Yakutia, Rodoman notes, an area the size of Switzerland has been disfigured), and state support for uncompetitive production. Granted, rolling up inefficient and harmful activity comes at the price of swelling the numbers of "economically redundant" Russian citizens. But for that, in theory, rich countries ought to pay, so as not to lose the forest riches of Siberia and preserve as part of the global common wealth, the purity of the Sayan and Altai Mountains or Lake Baikal.

Sociological research declares that outside our cities a multi-million person army of pension and welfare recipients lives year round. This public dole must be increased to support rural resettlement. After all, if millions of Soviet people once received a miserly paycheck from the government for then-useless pseudo-labor (recall the sad joke of those years: We pretend to work, you pretend to pay us), then why couldn't their children and grandchildren receive from other governments a fair reward for refraining from harmful ecophobic activity? At the same time, a life on welfare is far from always the same as parasitism and idleness. Just the opposite, with the right frame of mind, it allows a person to find their way to the kind of activity they are truly meant for, whether it is communing with nature and actively caring for it, looking after children, household management, artistic creation or starting a small business.

From the other end, residents of Russia's major cities would have a healthier natural environment for everyday life and creative activity. Active mental labor not only combines beautifully with ecological tourism, it could not exist without it. For good reason were the most ardent fans of tourism in the USSR engineers and scientific workers from academic institutions and the Military-Industrial Complex who intuitively found a curative outlet for themselves in mountain climbing, canoe trips along the rapids and other forms of extreme leisure, along with the usual excursions into nature on weekends and holidays.

It's hard to say what role tourism played in the realization of our space program or in establishing the Soviet nuclear umbrella, but it is beyond doubt that it could still serve our scientific and engineering thought. And while we have already lost our working class, engineering and scientific thought is still warm and can be born again...if we establish the right circumstances. And instead of organizing routine assembly line production—clothing and shoes, telephones and computers—in competition with China, Russia could focus on a skilled construction orientation, on experimental and low-circulation production, then selling our ready-to-introduce technical designs to China, India or Indonesia.

But if, let's say, the United States is in essence an urban civilization (Ecumenopolis), beginning its history three centuries ago with a blank slate, then Russia's rural roots go still deeper, no matter how far seven Soviet decades have uprooted them. To this, in part, speaks the irresistible pull of urbanites newfound and trueborn alike to acreages beyond the city lines. Indeed, this passion often turns to misfortune for nature, since the division of plots belongs, as a rule, to corrupt bureaucrats who approach this business God only knows how, but in any case without considering the interests of the surrounding landscape, often turning it into a single human anthill. And dacha colonies in the forest, by Vladimir Kagansky's estimate, ruin an area five to six times greater than they themselves occupy (Rodoman 2002).

And, meanwhile, without the traditional villages and landed estates, without the inimitable middle-Russian landscape located just above the line of sight that once inspired great Russian artists, the people, according to Rodoman, cannot remain as one. Developed countries know this danger well. Therefore, the widespread practice of government support for agriculture—that clearly anti-market policy which begets serious battles in the EU and WTO—takes place not only in the interest of providing food security, but for the preservation of a rural way of life that corresponds to modern ecological, technological and economic standards.

In all likelihood, much of what we have said above may seem like a dream beyond the scope of real life, an impossible utopia, and it's hard to find words to disagree. Anyway, many also view the idea of sustainable development as utopian. What looks utopian today, however, might tomorrow look like an unjustifiably missed opportunity. Such it has been more than once in history. But one thing seems beyond question: the future of Russia is unthinkable in separation from the countries of civilized Europe, and fate itself calls on them to complete and enrich one another. This fact, long obvious to many, should enter the consciousness of all the peoples living in that space.

References

Aksenov, D. E., Dobrynin, D. V., Dubinin, M. Y., et al. (2003). *Atlas of Russia's intact forest landscapes*. Moscow: Izdatel'stvo MSoES. Washington: World Resources Institute. Retrieved from http://old.forest.ru/eng/publications/intact/2.html.

Arsky, Y. M., Danilov-Danil'yan, V. I., Zalikhanov, M. C., Kondratyev, K. Y., Kotlyakov, V. M., & Losev, K. S. (1997). *Ecological problems: What's happening, Who is to blame, and What is to be done?* Moscow: MNEPU. [in Russian].

Boyd, I. L., Freer-Smith, P. H., Gilligan, C. A., & Godfray, H. C .J. (2013 November - 15). The Consequence of tree pests and diseases for ecosystem. *Science, 342* (6160).

CIA World Factbook. Retrieved from https://www.cia.gov/library/publications/the-world-factbook/index.html

Contribution of Russian Forests to the World Carbon Balance and Tasks of the Forestry Service After Ratification of the Kyoto Protocol. (2004). *Ustoychivoe lesopol'zovanie, 4*(6), 16–20. [in Russian].

Danilov-Danil'yan, V. I. (1999). *Sustainable development and problems of environmental policy*. Moscow: Ekosinform. [in Russian].

Danilov-Danil'yan, V. I. (2001). Russian ecology: Awaiting a breakthrough? In V. I. Danilov-Danil'yan (Ed.), *Race to the market: Ten years later*. MNEPU: Moscow. [in Russian].

Danilov-Danil'yan, V. I. (2006). The ecological significance of energy conservation. Energetika Rossii: problem i perspektivy. *Trudy Nauchnoy sessii RAN* (pp. 196–207). Moscow: Nauka. [in Russian].

Danilov-Danil'yan, V. I., & Losev, K. S. (2006). *Water usage: Ecological, economic, social and political aspects*. Moscow: Nauka. [in Russian].

FAO. (2010). *Global forest resources assessment: Key findings*. Rome: FAO.

Fomin, S. A. (2005). Main Government exectutive organs in the area of environmental management in Russia in 2005. *Rossiya v okruzhayushchem mire: 2005*. Moscow: Modus-K; Eterna.

Forest Encyclopedia. (1986). In two volumes. Volume 2. G. I. Vorobyov (Ed.). Moscow: Sovietskaya entsiklopedia. Retrieved from http://forest.geoman.ru/forest/item/f00/s02/e0002857/index.shtml [in Russian].

Friedli, H., Lotscher, H., Oeschger, H., Siegenthaler, U., & Stauffer, B. (1986). Ice core record of the 13C/12C-Racio in atmospheric CO2 in the past two centuries. *Nature, 324*, 237–238.

Global Forest Resources Assessment. (2005). 15 key findings. Retrieved from http://www.fao.org/forestry/foris/data/fra2005/kf/common/GlobalForestA4-ENsmall.pdf

Global Forest Resources Assessment. (2015). www.fao.org/3/a-i4793r.pdf

Gorshkov, V. G. (1995). *Physical and biological bases for sustainable life* (472 p). Moscow: VINITI [in Russian].

Gorshkov, V. G., Gorshkov, V. V., & Makarieva, A. M. (2000). *Biotic regulation of the environment: Key issue of global change.* London: Springer.

Gorskov, V. G., Kondratyev, R. Y., & Losev, K. S. (1998). Global eco-dynamics and sustainable development: scientific aspects and the "human measurement". *Ekologia, 3*, 163–170. [in Russian].

Hannah, L., Lohse, D., Hutchinson, C., Carr, J. L., & Lankerani, A. (1994). A preliminary inventory of human disturbance of world ecosystems. *Ambio, 23*, 246–250.

Hansen, M. C., Potapov, P. V., Moore, R., Hancher, M., Turubanova, A., Tyukavina, A., Thau, D., Stehman, S. V., Goetz, S. J., Loveland, T. R., Kommareddy, A., Egorov, A., Chini, L., Justice, C. O., & Townshend, J. R. G. (2013). High-resolution global maps of 21st-century forest cover change. *Science, 342*, 850–853.

Houghton, R. A. (1989). The long-term flux of carbon to the atmosphere from changes in land use. Extended abstracts of papers presented on the Third International Conference on analysis and evaluation at atmospheric CO2 data. Heidelberg: W.M.O. University. pp. 80–85.

Houghton, J. T., Meira Filho, L. G., Callander, B. A., Harris, N., & Kattenberg, A. (1996). Climate change 1995. In K. Maskels (Ed.), *The science of climate change.* Cambridge, Cambridge University Press.

Kondrat'yev, K. Y., Karapivin, V. F., & Potapov, I. I. (2005). Natural disaster statistics. In Environmental problems and natural resources: General information (pp. 57–76). Moscow. [in Russian].

Kondratyev, K. Y., Danilov-Danil'yan, V. I., Donchenko, V. K., & Losev, K. S. (1993). *Ecology and politics.* St. Petersburg: Sankt-Peterburgsky Naucho-Isledovatelny Tsentr ekologicheskoy bezopasnosti RAN. [in Russian].

Lorius, C., & Oescher, H. (1994). Paleo-perspectives: Reducing uncertainties in global change? *Ambio, 3*(1), 30–36.

Losev, K. S. (2001). *Ecological problems and prospects for sustainable development in Russia in the 21st century.* Moscow: Kosmosinform. [in Russian].

Losyev, K. S., Gorshkov, V. G., Kondratyev, K. Y., Kotlyakov, V. M., Zalikhanov, M. C., Danilov Danil'yan, V. I., Golubev, G. N., Gavrilov, I. T., Revyakin, V. S., & Grakovich, V. F. (1993). In V. I. Danilov Danil'yan & V. M. Kotlyakov (Eds.), *Problems of Russian ecology.* Moscow: VINITI. [in Russian].

Maksakovskiy, V. P. (2008). A geographical portait of the world (in two books). Moscow: DROFA. [in Russian]. Book 1. Retrieved from http://www.twirpx.com/file/997779/. Book 2. Retrieved from http://www.twirpx.com/file/997899/

Meadows, D. H., Meadows, D. L., & Randers, J. (1992). *Beyond the limits.* Post Mills: Chelsea Green.

Myneni, R. B., Dong, J., Tucker, C. J., Kaufman, R. K., Kauppi, P. E., Liski, J., Zhou, L., & Alekseev, V. (2001). A large carbon sink in woody biomass of Northern forests. *Proceedings of the National Academy of Sciences of the United States of America, 98*, 14784–14789.

Naymark, E. (2013). *World forest masses are gradually being studied.* Website "Elementy". Retrieved from http://elementy.ru/news/432137 [in Russian].

Nefyodov, A. V. (2008, Aug 25). "Those Russians, did they use to have rivers? On the question of building the Evenkiyskaya (Turukhanskaya) hydropower on the Nizhnyaya Tunguska river". *Turukhanskaya Shirota, 35.* Retrieved from http://www.bioticregulation.ru/life/life10.php

Olsson, R. (2009). *Boreal forests and climate change.* Goteborg: Air Pollution & Climate Secretariat & Taiga Rescue Network.

Raynaud, D., Jouzel, J., Barnola, J. M., Chapellaz, J., Delmas, D. J., & Lorius, C. (1993). The ice record of Greenhause gases. *Science, 59*, 926–934.

Rimashevskaya, N. M. (2003). In *Just and unjust social inequality in Modern Russia* (pp. 129–145). Moscow: Referendum. [in Russian].

Rodoman, B. B. (2002). The Great Landing (Paradoxes of Russian Suburbanization). *Otechestvennye zapisi, 6*(7), 404–416. [in Russian].

Rodoman, B. B. (2004). Russia in an Administrative-Territorial Monster November 4, 2004. POLIT.RU. Retrieved from http://polit.ru/article/2004/11/04/rodoman/ [in Russian].

Rodoman, B. B. (2012). Russia's ecological specialization. INTELROS—Intelektual'naya Rossia. Retrieved from http://www.intelros.ru/subject/figures/boris-rodoman/12628-ekologicheskaya-specializaciya-rossii.html [in Russian].

Solzenitsyn, A. I. (1990). *Kak nam obustroit Rossiyu? Posil'nye coobrazheniya (Rebuilding Russia: Reflections and Tentative Proposals)*. Leningrad: Sovetskiy Pisatel'.

Staffelbach, T., Stauffer, B., Sigg, A., & Oeschger, H. (1991). CO2 measurements from polar ice cores: more data from different sites. *Tellus, 43B*(2), 91–96.

StatBureau.org. (2007–2015). Retrieved from https://www.statbureau.org/ru/united-states/inflation-tables

State of the World's Forests 2012. (2012). Rome: FAO. Retrieved from http://www.fao.org/3/a-i3010e.pdf

TEEB. (2010). *The economics of ecosystems and biodiversity: Mainstreaming the economics of nature: A synthesis of the approach, conclusions and recommendations of TEEB*. Retrieved from http://www.biodiversity.ru/programs/international/teeb/materials_teeb/TEEB_SynthReport_English.pdf

The Ecological Footprint Atlas. (2010). *Global footprint network*. Oakland.: Retrieved from http://www.footprintnetwork.org/content/images/uploads/Ecological_Footprint_Atlas_2010.pdf.

Vitousek, P. M., Ehrlich, P. R., Ehrlich, A. H. E., & Matson, P. A. (1986). Human appropriation of the product of photosynthesis. *Bioscience, 36*(5), 368–375.

Vompersky, S. E. (1994). The biospheric significance of Bogs in the carbon cycle. *Priroda, 7*, 44–55. [in Russian].

Watts, J. A. (1982). The carbon dioxide questions: data sampler. In W. C. Clark (Ed.), *Carbon dioxide review*. New York: Clarendon Press.

Wofsy, S. C., Goulden, M. L., Munger, J. W., Fan, S.-M., Bakwin, P. S., Daube, B. C., Bassow, S. L., & Bazzaz, F. A. (1993). Net exchange of CO2 in a mid-latitude forest. *Science, 260*(5112), 1314–1317.

Worldstat Info. (n.d.) Retrieved from http://en.worldstat.info/

Worldwatch Database. (2000). Washington: Worldwatch Institute.

Zalikhanov, M. C., Losyev, K. S., & Shelekhov, A. M. (2006). Natural ecosystems as the key natural resource of humanity. *Vestnik rossiiskoi akademii nauk., 76*(7), 612–614. [in Russian].

Chapter 16
What About Coevolution?

In both its scale and its significance for the fate of world civilization, the task of transitioning to sustainable development with the aim of averting catastrophe in the biosphere surpasses anything humanity has had to ever overcome. And though in the half century that has passed since global ecology arose as the stubborn object of world society's attention society has progressed materially in grasping the intellectual scope of the problem, but it has physically achieved nearly nothing. And even isolated, unquestionably positive shifts pale in comparison to the growing destructive influence the environment has suffered over the same period. We cannot name that anything but running in place, or more like a retreat and surrendering of positions across a number of important ecological fronts. And here these five decades, counting from the mid-1960s, however brief a moment they might seem in contrast to geological or even historical epochs, are clearly a timeframe of the magnitude that separates us from the beginning of irreversible change in the biosphere, if, we might hope, it has not yet begun.

And though the issue of sustainable development as before attracts attention from the most various fields of knowledge, the rift between the speeds at which its scientific basis and philosophical and methodological superstructure are formed cannot help but arouse worry. The theoretical superstructure has arisen at an accelerated pace in recent decades, over a deficit of positive, concrete knowledge, which has led to an incompatibility of content used in the basic concepts of different authors. We can apply all this to such currently "fashionable" terms as a *coevolution*, a mutualistic development of nature and society, the *noosphere*. Here, for example, is the importance Nikita Moiseyev places upon these concepts in one of his last writings:

"The term 'noosphere' has at present received sufficiently widespread usage, but is interpreted very differently by various authors. Therefore, in the late 1960s, I began using the term 'epoch of the noosphere.' That was what I called the stage in the history of man (anthropogenesis, if you will) when his collective mind and collective will became able to provide the joint development (coevolution) of nature and society. Humanity is a part of the biosphere, and realizing the principle of

© Springer International Publishing AG, part of Springer Nature 2018
V. I. Danilov-Danil'yan, I. E. Reyf, *The Biosphere and Civilization: In the Throes of a Global Crisis*, https://doi.org/10.1007/978-3-319-67193-2_16

coevolution is a necessary condition to provide for its future… In Rio de Janiero, an attempt was made to formulate a certain general position, a general scheme for the behavior of the planetary community which took the name sustainable development… I saw it as wiser to think/consider it identical to the term "coevolution of man and biosphere." That is why I consider the development of a sustainable development strategy a particular step in the epoch of the noosphere, i.e. a step on the path of noospherogenesis" (Moiseyev 1997).

We immediately notice that the expression "epoch of the noosphere" is a bit less clear in meaning than "noosphere" itself. After all, we find no criteria for "noosphericality" either in literature or from Moiseyev himself. With regard to coevolution, is there any need to duplicate a concept if it is identical to the term sustainable development, which in translation to local languages is used by the whole world? And things aren't so simple with calling them identical, either. Truly, equating coevolution with sustainable development or one of its modifications is not something that comes up only in Moiseyev. There have been several works published on the topic in Russia alone that hold practically the same position (Rodin 1991; Karpinskaya et al. 1995, etc.).

Meanwhile, in its initial usage, the concept of coevolution only meant mutual adaptation or corresponding changes in species over the course of evolutionary biological development. Soon, however, it became clear that the word reflected a broader sphere of phenomena connected to the evolution of any mutually adaptive systems or elements of a single system in which the experienced changes do not create a mutually negative influence. In this way, for example, Eugene Odum (1983) determined nine types of interaction between biological populations which we may more or less justifiably view as variants of coevolution.

Analysis of coevolution in nature and society, however, is a complex and specific task that requires a special approach. At that you must not forget the most important aim in this case—to resolve the ecological crisis by way of transition to sustainable development—in relation to which coevolution presents itself either as a means to its realization or even as a substitute concept.

So how should we understand the coevolution of nature and society (of biosphere and man)? The answer to that question depends on your view of the inter-relation between the "coevolving pair." And in that sense, the widely held formula used by Moiseyev, "humanity is a part of the Biosphere," could hardly meet with any disagreement. Along with this, the fundamental asymmetry of relations in coevolving systems, in this case, humans and the biosphere, calls doubt on the very validity of posing the question this way. We are, after all, talking about coevolution of a part and the whole, and a part and the whole are fundamentally asymmetrical. But let us say there is still such a possibility. Then we must make a small concession for the clarification of some concepts. First, the concept of the biosphere (and its evolution), and, second, the evolution of humans (society).

The classical systems understanding of the medium leaves room for only one satisfactory definition of the biosphere: a system that includes the biota (the sum of all living organisms, humans among them) and their surrounding environment (the sum of objects under the influence of the biota and/or influencing it.

In the given context not just any influence interests us, but first of all that which might be material in the fate of civilization and the survival of humans as a species. It is at that systems coordinate that evaluations of change take on meaning: desirable or undesirable, tolerable or intolerable.

With concern to the biosphere's evolution, recalling the role played by living organisms in forming the ocean, atmosphere, soil and rock, we must give the most important place to the evolution of the biota, which occurred by means of speciation. At the same time, due to the systems character of the biota, the appearance or disappearance of any species from the arena of life inevitably sets off a wave of species changes in the ecosystems one or another species was "assigned" to. The speed of this process is determined by the time a species exists (about 3.5 million years on average) and the term of its formation (by current estimates, on the order of 10,000 years). Though there are also grounds to suppose that temporary intermediate characteristics have remained unchanged over the course of at least several hundred million years (Danilov-Danil'yan 1998).

When we look to the evolution of human society, we see that it obeys entirely different rules and unfolds against the background of the genetically unchanged constant of the species *Homo Sapiens* by way of developing social structures, social consciousness, material and spiritual culture as well as productive, scientific and technological potential. The most interesting aspect for us, however, will be humanity's growing influence on the biosphere over the course of this evolution. In the past two or three centuries, this has been determined primarily by the speed of scientific progress, or techno-evolution. And since innovation fuels this process, which in some respects resembles speciation, it may be helpful to compare the relative speed of both one and the other.

The thing is, material production, like the biota, organizes along systemic but randomly arising lines. And, as a rule, any innovation, the appearance of any new technological element in the sphere of production or management, brings about a wave of other innovations in the corresponding "technological niche." But while the pace of biological evolution remains almost constant for tens of millions of years, the speed of techno-evolution grows incessantly. By the end of the twentieth century, for example, the innovation cycle in advanced industries took, on average, about 10 years.

Now, compare these two numbers: 10 years are required to create new productive technologies, and 10^4 years are spent to form new species, new "natural technologies." Is it valid at a difference of three orders of magnitude to speak of the possibility of some kind of cooperative evolution of nature and humanity? And if valid, in what form should the biosphere arrange co-evolutionary changes in response to innovations in human activity? Perhaps new species should arise in time to adapt to the pace of anthropogenic influence? Let's say a new species of bacteria turns up, able to decay plastics or turn mountains of used aluminum cans into bauxite and nepheline. That kind of co-evolution would probably come in handy, but the very absurdity of such a proposition tells you all you need to know.

Of course, human influence on the biota doesn't go entirely unanswered, including by way of speciation. But due to the enormous difference in speed between bio-evolution and techno-evolution, this reaction simply doesn't have time to bear fruit. Technogenic influences replace each other before the response becomes noticeable.

Perhaps mankind someday will find a way to speed up speciation in the biota, thus increasing its "co-evolutionary capabilities," by way of artificially creating new species or genetically modifying existing ones? Not even discussing the most dangerous consequences of introducing organisms with an artificial genetic structure into nature, let us just say that the realization of such plans would mean the end of the biota's natural evolution and its transformation into a system whose development is directly regulated by man. But in that case, is it even worth discussing the co-evolution of the biosphere and humans? It would be almost the same as talking about the "coevolution" of regularly updated automobiles and the people who drive them, even if the former occasionally gets out of the latter's control.

***

The results of human influence on the environment have undergone scientific analysis several times in recent decades. And, as data from many observations and research papers testify, practically all human activity from the moment we mastered fire and transitioned from hunting and gathering to herding and farming has meant one thing: disruption. And the reaction of any system to disruption depends primarily upon its scale, that is whether or not the disruption has surpassed the acceptable threshold. Meanwhile, the system's capacity for self-restoration, as you know, is not unlimited. It preserves itself only up to a certain critical point, after which irreversible processes unfold. These will either destroy the system entirely or fundamentally change its structure so that it, on being reborn, becomes a different quality (Danilov-Danil-yan 1998).

It stands to reason that the biosphere, like any highly organized system, also has its threshold of sustainability. But how does one determine where the boundary of acceptable change lies, and which parameters could be key in this sense? And how does one prognosticate which subsystems (species, organism communities) will hold up in the event of transformation beyond the threshold, and which cannot survive the transformation? After all, when humans provoke the death of a species, it is to one extent or another an event which disorganizes the biosphere, and who can predict where this wave of disorganization will stop or how many species and their communities may be swept from the arena of life? And will it not, as we might guess from other such waves, spread to the biosphere as a whole? To look at another side of the problem, any reaction to external disruption requires a certain amount of time (delay effect). Who can say if we have not already come to the threshold where the biosphere's reaction to previous disruptions, combined with new ones, makes hopeless any attempt to link cause and effect or understand the meaning of any one anthropogenic influence?

We don't know the answer to these or any such questions. But only unthinking naivete could hold out hope that everything will work out on its own, or that the train humanity must take to sustainable development will patiently wait for us at "Ecology Station." How long do you think it can stand there idling?

We do have a still very influential group of "technological optimists" who suppose that scientific progress will have the power to solve any problem born of civilization, including the ecological ones. But in that case, what can we expect from technologically armed humanity? Obviously, it will take for itself some functions or other to maintain environmental sustainability. Indeed, we could introduce examples of when people have succeeded in artificially cobbling over the gaps in the "workshop of nature" that they themselves brought about. One example is the International Birding & Research Centre—Eilat (IBRCE), where millions of avians have recently found refuge—predators, songbirds, waterfowl, representatives of dozens of bird species that migrate each year from Eurasia to Africa and back.

The center is located on the shores of the Red Sea, in Israel, where it meets the Gulf of Aqaba on a key territory of one of the world's three migratory corridors. The birds fly in exhausted from a multi-day flight over the Sahara and Red Sea. There is no other place for rest and recuperation: to the east stretch thousands of miles of Arabian Desert, and ahead on their route lie the Negev and Dead Sea where they will find nothing to eat. But in the 1960s, people built an oil pipeline on the spot with an oil storage facility, and later a resort was unveiled, thus depriving the birds of their only available rest stop. Fortunately, however, people thought better of it in time, and in 1993, on the spot of an overfilled dump, at the initiative of famous Israeli ornithologist Reuven Yosef and vehemently opposed by local hotel operators, one of the largest ornithological stations in the world was founded. They cut open sources for fresh water, fenced off the territory from outside intruders, and, most importantly, supplied food for the avian lodgers. In this way, the Eilat Center became not only a place to study and tag birds, but also a type of hotel for birds, or perhaps a refueling station. Millions of dollars have been spent on its maintenance so far and one might confidently assert that one the efforts of this small, self-sacrificing team in many ways depends the biological equilibrium of enormous swaths of the Paleoarctic, from Scotland to the Black Sea Steppes (Yosef 1996, 2002).

But what exactly does this example tell us? That, figuratively speaking, when the beams of our planetary home begin to cave in, in a number of cases, humans are capable of propping them back up. In such a home we might live for a time, putting up girders here and there so long as the structure does not finally collapse. With regard to the idea of a managed environment, as we saw in Chap. 12, it looks completely utopian. Today, at least, humans do not have the technological means at their disposal, or the scientific R&D, that would allow them to establish artificial regulation of the environment. And what's more, there is not a single example of environmental protection achieved through the application of new technologies where local environmental improvement was not achieved at the cost of worsening the overall ecological balance. After all, the technologies used in these cases are inevitably linked to energy expenditure, the cost of which, borne by the global environment, exceed the sum local benefit.[1] And even if we concede the hypothetical possibility

---

[1] Examples of destroyed ecosystem restoration, such as putting down forests in place of tree farms, clearing obstructions from springs, etc., naturally do not belong to this category, since they are based on the application of natural, biotic means rather than technological ones. These more likely

of some radical breakthrough in the future, the outlook is no less cloudy all the same, while the threat of the ecological crisis spiraling into a biospheric catastrophe is so palpable and real, the people simply do not have time for the enactment of such ambitious projects.

In this way, from whatever direction you approach it, whether ecological or technological, one does not find any convincing basis for posing the question of cooperative development of the biosphere and society. And what's more, we cannot speak of any possible evolution of the biosphere "in a human direction." That, however, does not exclude the opposite: human evolution "in the direction of the biosphere" with a gradual weakening of anthropogenic pressure and a review of some basic tenets of modern civilization. That is the essence and spirit of sustainable development, which is humanity's only chance to prevent biospheric catastrophe and find haven in a more or less stable future. And if they rid themselves of a hypnosis of the false significance hiding behind a façade of certain terms, what alternative could the supporters of the idea of coevolution offer to this one possible course for civilization?

***

While coevolution as a concept arose relatively recently, another more popular term, noosphere—the sphere of reason, the "envelope of thought" around the Planet—was already in use by the 1920s. The term's origins begin with two French "Bergsonians," mathemetition and philosopher Edouard Le Roy (1870–1954) and anthropologist and philosopher Pierre Teilhard de Chardin (1881–1955). Of particular note, however, is the contribution Vladimir Vernadsky made to the birth of this idea.

In 1922, Vernadsky received an invitation from the Sorbonne to read a course of lectures on geochemistry, a new science at the time which he was actively working to develop. But Vernadsky arrived in Paris with a store of fresh ideas that went beyond the field of geochemistry. He brought in his intellectual suitcase his formulated concept of the biosphere as an enveloping layer of the Earth, uniting within itself the living and non-living matter with the former playing the leading role in making up the face of the planet. He also called attention to the unique place that humans occupied in these processes, forming, in his words, an independent geological force.

Before meeting Vernadsky, neither LeRoy nor Teilhard de Chardin had obtained data concerning human influence on the biosphere since they had no awareness of the concept. Thus Vernadsky's four-year sojourn in Paris and lectures read at the College de France, where LeRoy and Teilhard de Chardin sat among the audience, as well as a speech at Henri Bergson's seminar (unfortunately, no meeting took place between the two men) provided great benefit to both sides. As a result, LeRoy and Teilhard de Chardin got the opportunity to put a broad, hard, scientific foundation under their concept of the noosphere, and Vernadsky, putting his French colleagues' idea in his arsenal, laid out his view of the noosphere as a stage in the

---

demonstrate to us the possibilities of "natural reconstruction" through the power of nature itself, in cases where humans abandon natural objects instead of engaging in ecophobic activities.

biosphere's development. That is, it made a decisive contribution to the establishment of his school of thought. In the late 1930s he wrote, "Humanity, taken as a whole, has become a powerful geological force. And before it, before its thought and labor, stands the question of reconstructing the biosphere in the interests of a free thinking humanity as a single whole..." (Vernadsky 1993, p. 305). "Under the influence of scientific thought and human labor, the biosphere is transitioning into a new state—into the noosphere" (Vernadsky 1997).

As we have already noted, this was a time of historical optimism, faith in the bright future ordained for humanity. Despite the lengthening shadow of fascism over Europe, despite the horrible sacrifices borne through the First World War and over the course of the "Communist experiment" in the Soviet Union, Vernadsky, like many of his Western colleagues, linked this future with the unlimited possibilities of human reason and thought that its prerequisites had already been laid in modern civilization "...The civilization of cultured humanity," he wrote, "to the extent it is a form of organization for a new geological force established in the biosphere, cannot be cut off and destroyed as it is a great natural phenomenon corresponding historically, or more accurately, geologically, to the apparent organization of the biosphere. In creating the noosphere, it links itself with all its roots to the envelope of the earth, in ways that earlier in the history of humanity did not exist to any comparable extent (Vernadsky 1988, p. 46).

Along with this, neither Vernadsky, nor LeRoy, nor Teilhard de Chardin yet doubted that human influence upon the biosphere was beneficial on the whole, though all thought that this spontaneous process should be directed into some rational avenue. Affirming that this influence should be subject to management, and that the spontaneity would in time give way to conscious direction, Vernadsky did not foresee any danger in the very scale of influence and did not see a threat from that end. He put his greatest hopes in science, which would open to humanity previously unheard of possibilities. "Such a sum of human actions and ideas," he wrote, "has never been before, and it is clear that this movement cannot be stopped. In part, before scientists in the near future will stand the entirely new tasks of conscious direction of the noosphere's organization which they cannot avoid since the spontaneous course of scientific knowledge's growth is pointing them that way" (Vernadsky 1988, p. 50).

In this way, Vernadsky understood the noosphere as a kind of step in the development of nature and society, when man, armed with scientific knowledge and as the only species to have achieved supremacy above all others, takes upon himself all responsibility for the Earth's biosphere, reforming and reshaping it in accordance with the laws of nature he has discovered. That is, according to Vernadsky, the spontaneity of development gives way to consciously planned changes to the environment, and in place of a chaotic assortment of various conflicting nations and peoples a single, rationally organized humanity would enter the stage, armed to the teeth with knowledge, technology and acquired historical experience.

Of course, the decades gone by since the founder of the biosphere school of thought passed away have sharply changed the face of the world around us, but nonetheless Vernadsky correctly guessed some then barely noticeable tendencies.

For example, the process of globalization occurring before our eyes, connecting every corner of the earth by television, internet, mobile and satellite link into a single information space, sounds like Vernadsky's prediction of "instant transfer of thought, its simultaneous discussion on the whole Planet" (Vernadsky 1997). We cannot help but pay our respects to his ingenious far-sightedness, with which he, in the age of general political disorder and confrontation of the 1930s, was able to come to the conclusion that humanity's further spontaneous development was impossible, and of the responsibility laid upon it for the fate of the biosphere.

And yet the past half century has shifted some of the emphasis, and much, from the towering heights of today, looks different. It is already hard for us to share Vernadsky's optimism with regard to the limitless possibilities of scientific progress, which has turned out, upon trial, to be helpless in solving a number of problems beyond the strength of a matured humanity that caused many of them to begin with. Along with this, the enormous breakthroughs in life and earth sciences over recent decades have unbelievably expanded our information on the biosphere, as a result of which we know much more about it today than would have been available to Vernadsky. In part we have a much better image of the whole complexity of the mechanism which provides environmental stability on our planet, and next to which humans, powerfully armed with the most innovative technology, appear far more humble than they did in Vernadsky's day.

Due to this, his thesis of the biosphere's transition to the qualitatively new state of the noosphere arouses great doubts. The transition of the biosphere to any different state, after all, is possible only in the case of its extreme disruption, in consequence of which it will lose its capability for self-restoration. Such a disruptive force took place, for example, in the great Ice Ages, when a return to a new level of stability was reached by way of radical reconstruction of the biota's internal structure, stretched out over a period of millions of years. Modern humanity's technological possibilities could also lead to disruption of the biosphere beyond its limits and the irreversible loss of its stability, but that would be much worse, equivalent to an ecological catastrophe. By the way, even in this case, the biota is theoretically able to return the biosphere to a new level of sustainability, which again would require millions of years. But such an outlook hardly represents any interest for humanity. After all, in that radically reconstituted biosphere, there would already be no place for us, or for the majority of other modern species. And, naturally, that course of unfolding events is not what the creators of the noosphere concept had in mind when speaking of the biosphere's transition, under the influence of human civilization, to a fundamentally new state.

But perhaps our understanding of the noosphere could be applied to that future, undisturbed state of the biosphere toward which the idea of sustainable development orients itself? Then we will try putting the question differently: could the biosphere transition to any other quality at all, besides the one it has come to over billions of years through the process of unending evolutionary endeavor? And what are humans able to give it, aside from reducing the disruption of anthropogenic influence that has ended only in misery for the biosphere? We think that throughout this entire book, we have attempted to answer that question. And, therefore, thinking of some different biosphere formed on the basis of ultramodern technologies or even those yet unborn, we are in essence saying, like Chekhov's character, "It cannot be, because it could never be." And, all the more, the concept of the noosphere cannot be applied to the present, disturbed state of the environment, fraught with irreversible degradation. In other words, we cannot apply it to any or all possible states of the biosphere.

As well the idea of the noosphere somewhat opposes the image of a future world order with common human, humane values and a conception of a new level of development for society, in Pushkin's words, "when people, forgetting their squabbles, unite into a single family," in order to transition from a spontaneous development of civilization to a forecasted and conscious choice of our further path. Probably, the interpretation of the noosphere as a particular state of human society, developed in recent days by several ecologists and philosophers (Danilov-Danil'yan 1998; Ursul 1998) is closer to Teilhard de Chardin's approach than that of Vernadsky and LeRoy, Though it is better to leave out Teilhard de Chardin's mystical "Omega Piont," which lies beyond the bounds of understanding the problem scientifically. One might envision the achievement of such a "noospheric state" as a form of natural and social harmony in which human thought understands the limits of human action laid upon it by the environment, and chooses the development path safest for the future.

References

Danilov-Danil'yan, V. I. (1998). Is "Coevolution" of nature and society possible? *Voprosy filosofii, 8*, 15–25. [in Russian].

Karpinskaya, P. S., Liseyev, I. K., & Ogurtsov, A. P. (1995). *Philosophy of nature: Coevolution strategy*. Moscow: Interpraks. [in Russian].

Moiseyev, N. N. (1997). Coevolution of nature and society. The path to noospherogenesis. Ekologia i zhizn 2/3. Retrieved from http://www.ecolife.ru/jornal/echo/1997-2-1.shtml [in Russian].

Odum, E. (1983). *Basic ecology* (p. 518). Philadelphia: Saunders.

Rodin, S. N. (1991). *The idea of coevolution*. Novosibirsk: Nauka. [in Russian].

Ursul, A. D. (1998). *Russia's transition to sustainable development: The noosphere strategy*. Moscow: Noosphera. [in Russian].

Vernadsky, V. I. (1988). *Philoshophical thoughts of a naturalist*. Moscow: Nauka. [in Russian].

Vernadsky, V. I. (1993). A few words on the noosphere. In *Russian cosmism: An anthology of philosophical thought*. Moscow: Pedagogika-Press. [in Russian].

Vernadsky, V. I. (1997). *Scientific thought as a planetary phenomenon*. (B.A. Starostin Trans.). Moscow: Nongovernmental ecological V.I. Vernadsky Foundation.

Yosef, R. (1996). Eilat, Israel: Avian crossroads of the old world. *Living Bird, 15*, 22–29.

Yosef, R. (2002). *Pollution in a promised land—an environmental history of Israel*. Berkeley and Los Angeles: University of California Press.

Conclusion: "The Die is not yet Cast"

Most likely the reader, having read this book to its end, feels the question involuntarily arise: well, what next? The uncertainty of the future oppresses you and leads you to gloomy thoughts. That, anyway, is what always happens at the turning of ages, such as 1917 in Russia, for example, or in France at the end of the eighteenth century. When society goes mad, when our accustomed way of thinking and basic principles undergo a test of strength, no one can guess which card may fall from the deck of history. And then much will depend on the confluence of circumstances. Had not, for example, Napoleon turned up in the heat of a power struggle in October 1799, or Lenin and Trotsky after the February Revolution in Petrograd, the histories of France and Russia could have gone along entirely different paths. Meanwhile, in Victorian England or in Russia under Nicolas I, life went on as scheduled for decades, little depending upon the actions of any given person.

Today we live at the turning of an age, under conditions of high disequilibrium: disequilibrium in nature; disequilibrium in society. But here we are not speaking of any one country or any group of countries, but of the human ecumene as a whole, and this instability may tell in the most unpredictable ways upon the fate of humanity.

Many have written about how society is not keeping up with the pace of changes brought about by human progress. You've read about them in the pages of this book as well. But, in recent times, science has been enriched by new conceptions of the instability of complex systems, including those in society and nature, which we owe primarily to research in the areas of thermodynamics and non-equilibrium physics. These conceptions cast doubt upon the paradigm of predetermined human development (as Lev Tolstoy thought of it, for example), particularly in periods of crisis when becomes palpable the dependence on isolated, occasionally random events, or fluctuations in the terms of hard science.

"Events are dust," noted French historian Fernand Brandel. Disagreeing, Nobel Laureate Ilya Prigogine, who made a particularly important contribution to the development of thermodynamics and dissipative structures theory (synergetics), wrote: "What is an event? An analogy with 'bifurcations', which are studied above

© Springer International Publishing AG, part of Springer Nature 2018
V. I. Danilov-Danil'yan, I. E. Reyf, *The Biosphere and Civilization: In the Throes of a Global Crisis*, https://doi.org/10.1007/978-3-319-67193-2

all in non-equilibrium physics, comes immediately to mind. These bifurcations appear at special points where the trajectory followed by a system subdivides into 'branches'. All branches are possible, but only one of them will be taken. One does not generally see a single bifurcation; in general, a succession of them appear." He goes on, "The sciences of complexity therefore lead to a metaphor that can be applied to society: an event is the appearance of a new social structure following a bifurcation; fluctuations are the outcome of individual actions" (Prigogine 1999).

Here, how could we not recall certain particularly loud "fluctuations" from recent history? The collapse of the USSR, provoked by the August Putsch of 1991 and, like a domino effect, bringing about a whole cascade of "bifurcation" in the newfound independence of the republics, with bloody interethnic conflicts and streams of refugees. Or the late referendum in the UK, voting to leave the European Union ("Brexit"). That move was, in fact, initiated by a small group of people from the cabinet of Prime Minister David Cameron, clearly committing a dire mistake and not expecting such a result. Without speaking of how the referendum divided the nation in two, it may drag behind it a chain of very serious political, social and economic complications, both within Britain (Scotland declaring independence, for example) and without (right up to the collapse of the EU).

What could have brought about the referendum's scandalous outcome? The UK, after all, is a country with one of the highest educated populations, famed for its adherence to conservative traditions and not inclined to populism or spontaneous decisions. From what we can see, there occurred a rejection of certain aspects of globalization which least benefitted groups of the population such as pensioners and employees of the industrial and agricultural sectors (workers and peasants, one Russian journalist joked on that score). For good reason was it among these ranks that the idea of leaving the EU found its greatest support. And, meanwhile, to quote Prigogine again, "The task is to find the narrow way between globalization and the preservation of cultural pluralism, between violence and politics, and between a culture of war and one of reason" (Prigogine 1999). In other words, so that as few social groups, countries and peoples as possible feel aggrieved as a result of globalization. Right now more than enough of them do, and that is one of the most important destabilizing factors in the modern world. One more cause of instability, in Prigogine's opinion, is hidden in the feverish progress of information technology. Of course, the information society based on it ("society with a web structure") has unquestionable advantages, especially as part of economic infrastructure and medicine. "But there is information and disinformation; how can one tell the difference? Clearly, this requires ever more knowledge and a developed critical sense. The true must be distinguished from the false, the possible from the impossible" (Prigogine 1999).

A critical sense…That, we must say, is one of the key phrases that allows us to understand and evaluate much in the modern world. Let us go back once more to the unfortunate British referendum. It would seem that in a developed democracy with an independent media the population would have every possibility to impartially judge everything that is delivered to it from the TV screen, from the newspaper pages, in activist pamphlets or at rallies. It is notable, however, that it was mainly people with lower levels of educational attainment that voted for leaving the EU, while the youth, active users of the internet, were able to adequately evaluate the plusses and minuses of the proposed alternatives and voted against Brexit. What then can we say about countries with authoritarian regimes, where the majority of people, in the words of Russian humor writer Viktor Shenderovich, read one and the same book (if they are able to read at all) and watch one and the same state-run channels.

Unfortunately, in the cited article, symbolically named "The Die is not Cast," Prigogine does not touch on the crisis state of the environment. But there as well, obviously, much depends on the depth at which broad layers of the population recognize the problem. In any case, information about the real state of things ought to be brought forth through the education system to each family and each individual. And that, probably, is how we can oppose the threat of slipping into global ecological catastrophe. And though, as follows from complex system behavior theory, at turning points in development it is impossible to predict which direction the further course of events will go depending on a series of random fluctuations, the growing role of the human factor—individual and group decisions, actions and initiatives—forces us to look differently upon the place that ordinary people occupy in this process. In other words, on the measure of responsibility that lies on each of us and which must stamp out this "herd mentality," to use Stalin's phrase.

Prigogine put forward the quintessence of his essay in the following words: "My message to future generations is, therefore, that the die has not been cast, and that the branch taken following the bifurcation has yet to be chosen... But what will be the result of this bifurcation, along which branch of it are we going to find ourselves?" (Prigogine 1999) That was written in the year 1999, and we, unfortunately, do not know if the author would repeat it all today.

Reference

Prigogine, I. (1999). In F. Mayor (Ed.), *Letters to future generations*. Paris: UNESCO.

Index

Printed in the United States
By Bookmasters